THE NEW HEALTH BIOECONOMY

The New Health Bioeconomy

R&D Policy and Innovation for the Twenty-First Century

James Mittra

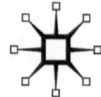

First published 2016 by
PALGRAVE MACMILLAN

The author has asserted their right to be identified as the author of this work in accordance with the Copyright, Designs and Patents Act 1988.

Palgrave Macmillan in the UK is an imprint of Macmillan Publishers Limited, registered in England, company number 785998, of Houndmills, Basingstoke, Hampshire, RG21 6XS.

Palgrave Macmillan in the US is a division of Nature America, Inc., One New York Plaza, Suite 4500, New York, NY 10004-1562.

Palgrave Macmillan is the global academic imprint of the above companies and has companies and representatives throughout the world.

Hardback ISBN: 978–1–137–43628–3
E-PUB ISBN: 978–1–137–43050–2
E-PDF ISBN: 978–1–137–43052–6
DOI: 10.1057/9781137430526

Distribution in the UK, Europe and the rest of the world is by Palgrave Macmillan®, a division of Macmillan Publishers Limited, registered in England, company number 785998, of Houndmills, Basingstoke, Hampshire RG21 6XS.

Library of Congress Cataloging-in-Publication Data

Mittra, James, author.
  The new health bioeconomy : R&D policy and innovation for the 21st century / James Mittra.
    p. ; cm.
  Includes bibliographical references and index.
  ISBN 978–1–137–43628–3

  I. Title.
  [DNLM: 1. Biomedical Research. 2. Biotechnology—economics.
3. Diffusion of Innovation. 4. Health Care Sector—trends.
5. Translational Medical Research. W 20.5]
R852
610.72′4—dc23   2015020483

A catalogue record for the book is available from the British Library.

For Braj Mittra (1936–2003)

Contents

Contents

Acknowledgments

The ideas contained in this book are the culmination of 12 years of research I have conducted within the Economic and Social Research Council's (ESRC) Centre for Social and Economic Research on Innovation in Genomics (Innogen)—now the Innogen Institute—at the University of Edinburgh. I would like to thank the ESRC, as well as the United Kingdom's Technology Strategy Board (TSB), in particular, for funding the various projects that generated the rich empirical data I use throughout this book. I would also like to give particular thanks to Joyce Tait and Dave Wield, who appointed me as an Innogen Research Fellow in 2003. They have both provided me with many years of support for my work and helped develop my academic career and intellectual interests in life sciences and the bioeconomy. I would also like to thank all those people who have collaborated with me on various projects, as well as those individuals who agreed to be interviewed, and whose anonymized accounts play an important role in illustrating my key arguments. I would also like to thank my Innogen colleagues, as well as many colleagues in Science, Technology and Innovation Studies (STIS) at Edinburgh, for the insightful discussions that have helped shape my academic interests over many years.

Chapter 3 of this book contains substantial material that was previously published in *Science and Technology Studies* 26 (3), 2013: 103–123. I would like to thank, and gratefully acknowledge, *Science and Technology Studies* for granting me permission to reuse this material.

Various people have commented on draft chapters of this book and I sincerely thank them for offering their valuable time, critical comments, and incredibly useful insights. They include Dave Wield, Niki Vermeulen, Isabel Fletcher, Fadhila Mazanderani, Geoff Banda, and Christine Knight. Special thanks are also due to my friend and colleague Nina Hallowell for reading the entire draft manuscript. I have tried to respond to all critical comments and suggestions, but any errors in the book are entirely my own. Finally, I want to thank my wife, Sara, for her continuing support and forbearance through the many postponed holidays and lost weekends as I endeavored to complete this book.

University of Edinburgh
April 2015

Chapter 1

New Biology and the Foundations of a Health Bioeconomy

Introduction

In 2004, scientists and entrepreneurs Craig Venter and Daniel Cohen, who together pioneered techniques to map the human genome, proclaimed that the twenty-first century would be defined by the biological sciences. They wrote: "While combustion, electricity and power defined scientific advance in the last century, the new biology of genome research...will define the rest" (Venter and Cohen, 2004: 73). Although reflecting a degree of hubris, the "century of biology" aphorism captures a shift in science and industrial policy and the emergence of a new zeitgeist in the latter part of the twentieth century. Biology came to replace physics as the exemplar "big science." The physical sciences had dominated science policy and political discourse for most of the century, and popular culture was defined by their perceived hazards, risks, and opportunities in finely balanced geopolitical contexts. Of course, in the context of commercial innovation, chemistry was predominant throughout the twentieth century. Furthermore, information and communication technologies (ICTs) brought major social and commercial transformations in the latter decades of the century. Nevertheless, the advent of new biology, and significant advances in life science technologies,[1] heightened expectations of a revolution in health care.

However, the transformative potential of the life sciences, and a fledgling bioeconomy, was tempered by scientific, technological, social, and commercial challenges and uncertainties. This is evident in the context of health and health systems, where molecular biology first took root in the clinic in the 1960s through developments in cytogenetic testing (Hopkins, 2006). In the United Kingdom, cytogenetic testing built on a long and rich history of publicly funded basic research. From the 1970s to 1990s, sophisticated screening technologies and new diagnostic and therapeutic options for drug development began to emerge. The development of recombinant deoxyribonucleic acid (DNA) technologies[2] in the 1970s was a major breakthrough in biological approaches to therapy, which had been incrementally evolving throughout the second part of the twentieth century. These new, step-change innovations would eventually disrupt conventional therapeutic pathways and commercial research and development (R&D) strategies.

The array of innovative technologies and medicinal therapies that are emerging in the twenty-first century challenges our existing regulatory systems and established health-care pathways, reimbursement systems, and clinical practices. These technologies and therapies are also restructuring entire industrial sectors. In the cases of regenerative medicine (RM) and stratified medicine, for example, de novo business models and value chains[3] must be created because there is no existing route to market (Mastroeni et al., 2012; Mittra and Tait, 2012). For RM, conventional preclinical animal studies for safety and efficacy are often inadequate, and the "gold standard" of a three-stage, placebo-controlled, randomized clinical trial (RCT) system is not always appropriate (Mittra et al., 2015; Webster et al., 2011). Furthermore, contemporary health innovation demands much greater participation by patients and publics in research, which raises important issues around consent and third-party use of personal data (Haddow et al., 2007; Mittra, 2007a). The value ascribed to the patient experience and perspective, and the increasing role of politically active patient groups in therapeutic innovation, means that patients are now at the center rather than the margins of biomedical R&D and the health bioeconomy. Pricing, reimbursement, and cost-effectiveness of new medicinal therapies are also continuing challenges for institutions and organizations responsible for purchasing and delivering health-care services, with the patient again central to these concerns. It is within this context that the notion of a complex and multifaceted health bioeconomy has gained social and political traction.

It is important to recognize that the use of new technologies and therapies (in this book I use "therapy" to refer only to medicinal products) within the clinic affects early-stage innovation options and strategies. Contrary to popular representations, modern therapeutic innovation does not follow a simple, linear path (Tait and Williams, 1999).[4] Furthermore, the fact that risk and uncertainty are intrinsic features of life sciences R&D is a continuing challenge for those responsible for foresight and policy analysis on science and technology futures (Williams, 2006). The long-term economic and noneconomic value claims underpinning new promissory technologies and therapies also appear opaque in the early stages of the innovation lifecycle, which makes identifying and exploiting sustainable routes to market difficult. The health-related life sciences, and the bioeconomy within which they are constituted and shaped, raise a number of important questions about novelty and value, organization and management of interdisciplinary R&D, and the blurring of boundaries between the laboratory and the clinic and between public and commercial spheres.

The Concept of "Innovation Ecosystem"

This book is about the evolution of new biology and the health bioeconomy in the twenty-first century. Throughout, I investigate how the institutional and organizational landscape for health R&D within Europe and the United States has been transformed, and reciprocally shaped, by new science and technology options. I critically explore this institutional and organizational change in the broader context of perceived problems facing contemporary health innovation and expectant stakeholder narratives that have coalesced around the concept of "translation." Here, translation denotes new policies and practices aimed at bridging the laboratory and the clinic to generate the promised social, clinical, and commercial benefits from significant investments in life sciences (Kraft, 2004, 2013; Mittra and Milne, 2013). More specifically, my objective is to reflect on the range of actors, institutions, and organizations that are now integral elements of a complex and distributed "innovation ecosystem" (Adner and Kapoor, 2010; Durst and Poutanen, 2013; Mastroeni et al., 2012).[5] Crucially, I address the impact of these changes on R&D practices and notional ideas of value and worth that circulate within the health bioeconomy.

The concept of innovation ecosystem highlights the interdependencies between different actors and organizations that co-produce new scientific knowledge, technologies, and therapies. It also captures the attendant social and institutional innovations that are necessary for products to reach a market and generate various types of value and benefit. The concept is much broader than many conventional theories of innovation systems, because it encapsulates the all-important social, economic, commercial, and policy/regulatory drivers. Central to the success of an innovation ecosystem is diversity, resilience, and robustness. In terms of how I employ the concept in this book, the ecosystem comprises the basic science, the individual business models and value chains for specific technologies and therapies, regulation (including product regulation as well as health technology assessment and reimbursement), funding agencies, markets, and, of course, the patients who are the ultimate beneficiaries of new therapies. It is only by capturing all these systemic elements, and relevant stakeholder interactions, that we can begin to understand both the opportunities and the challenges facing new biology and its application to health.

Of course, as Papaioannou et al. (2009) rightly warn, the term "innovation ecosystem" should not be used uncritically. It can lead to reductionist and functionalist accounts if it is translated as a straight biological metaphor and disassociated from the concept of ecology. I hope to avoid any pitfalls in the use of the term by grounding the concept in case examples of particular innovations that have highly interdependent organizational and institutional linkages and value chains. I also avoid the tendency to present health-related innovation ecosystems as universal explanations for the evolution of knowledge and technology dynamics in the sector. The utility value of the eco-system approach, for the purpose of this book, lies in the fact that it encourages us to broaden our analytical scope and explore the range of actors that produce knowledge, technology, and therapy for the growing health bioeconomy. Furthermore, it compels us to take seri-ously the notion of dynamism in the system and rethink our notions of value and waste. In presenting a broad systemic analysis of the health innovation ecosystem, I unveil some of the more substantive implications new biology has for industry, science, medicine, and soci-ety. It is at the nexus of these stakeholder communities that a diverse range of expectations and values are being negotiated and contested as new technological and scientific opportunities emerge.

A key question to explore is how different actors and organiza-tions constituted within the health bioeconomy operate in the inter-disciplinary environments required to successfully "do health R&D" in the twenty-first century. There is no simple answer, because dif-ferent stakeholder communities and practitioners are driven by their own institutional logics and subjective understanding of the value and challenges of health care driven by new biology (Mittra, 2013). Throughout this book, I conceptualize the creation or enactment of value within health innovation in its broadest sense. This is to avoid reducing complex issues and health-care product development pro-cesses to the crudest of economic metrics, or alternatively to valorize only the social and ethical dimensions. The bioeconomy encapsulates many different types of value and valuation practices, or what some usefully refer to as "orders of worth" (Stark, 2009). These shape the evolution of the science and the strategies of those responsible for its progress. However, while the benefits and impacts of new biology on therapeutic innovation are often contested, there are a number of perceived challenges and opportunities in contemporary health R&D that have inspired significant policy shifts and changes in regulatory, commercial, and broader socioeconomic norms and practices. The promissory value that is now ascribed to the concept of "translation," particularly within the science and policy communities, exemplifies the tectonic shifts that are taking place within biomedicine. Together, these have materially affected the everyday practices of R&D. They have also defined the availability and scope of new therapeutic options within the clinic.

In the remainder of this chapter, I outline what I consider to be distinctive about new biology in the context of a set of contemporary health innovation challenges that have led to significant change in how R&D is organized and practiced. The substantive chapters then provide rich illustrations of different aspects of these transformations, from the perspectives of the various organizations, institutions, and actors that comprise the innovation ecosystem. I also review some of the critical social science literature on the bioeconomy, and the expectations around future value that are being contested and debated. This clarifies the conceptual and theoretical approach that is used to frame the empirical material I present later. The key question I address in this chapter is: *What, if anything, is distinctive about new biology and the health bioeconomy, and how do they challenge conventional systems of health innovation and the enactment of value?* At the end of the chapter, I briefly explain the empirical data sources used to inform my arguments and outline the overall structure of the book.

What, If Anything, Is Distinctive about New Biology?

Since the inception of the Human Genome Project (HGP)[6] in 1990, and the announcement that the full human genome had been successfully sequenced in 2003, medical research, clinical practice, and the structure of the biopharmaceutical industries have undergone profound change. This is evident in the context of how R&D is now funded and organized. Translational Medicine (TM), as I discuss later in the book, has emerged as a powerful narrative and organizing principle for meeting the various challenges facing conventional drug development. These challenges include high failure rates for new drugs (referred to as "attrition" by the pharmaceutical industry), low productivity, and a perceived knowledge and culture gap between the laboratory and the clinic that new molecular biology appears to throw into sharp relief (Mittra et al., 2011; Mittra and Milne, 2013).

However, what, if anything, is truly distinctive or exceptional about this new biology? Are we merely being captured by the hype and hubris of the scientists, industrialists, and those who popularize science through depictions of dystopian or utopian futures when we glibly accept the new century as one defined by advances in molecular biology? Is the science of life ontologically and epistemologically distinct from other sciences? In his book, *The Disorder of Things: Metaphysical Foundations of the Disunity of Science* (1995), John Dupré makes the important point that when we look at science through the lens of contemporary biology, it appears far more disunified than sciences such as physics and chemistry, as they were portrayed by many philosophers of science in the first half of the twentieth century. These were the emblematic "big sciences" of the nineteenth and twentieth centuries. Biology, unlike physics and chemistry, does not exhibit the methodological unity or grand theoretical narrative to link the disparate fields under its domain. It therefore lacks a sense of internal disciplinary integrity. Furthermore, as Bensaude-Vincent (2007) rightly argues, biological systems cannot simply be reduced to a central code or program, as the crude, and false, analogy of the "gene as script" implies. Reductionism also makes little sense in the context of biological taxonomy, so biology, as presented by both Dupré and Bensaude-Vincent, is highly pluralistic in character. I argue that a distinctive feature of new biology is that it is necessarily interdisciplinary in nature and pluralistic in terms of how it is organized and managed. The implication for scientific practice is that people can work in various domains of the broadly defined "life sciences" without sharing strict methodological or epistemological commitments. Furthermore, they do not need to have unified objectives, or shared values and valuation practices, in terms of how potential application areas are prioritized and routes to market or clinic managed. I would further argue that the decentralized nature of the funding streams for the new biological sciences, and highly distributed innovation ecosystem within which specific application areas are shaped, contributes to, or at least reflects, this lack of disciplinary integrity and unity in knowledge production. It also generates uncertainty about product development pathways and the realization of value.

For the purpose of this book, I refer to both new biology and the life sciences as a specific set of interdisciplinary approaches, technologies, and scientific knowledge and expertise (tacit and codified) where "life" at the molecular level is a key component driving innovation and clinical practice. When *combined* with a particular set of organizational and institutional arrangements for "doing R&D," the novelty of the late-twentieth-century and early-twenty-first century biological sciences is illuminated. In the context of health, this structure is highly distributed, displays nonlinear attributes, and has the dynamics of an ecosystem. Of course, it is important to recognize that the foundations of new biology were being prepared long before the initiation of the HGP in the 1990s. From the 1970s, recombinant proteins were being developed for therapeutic use. Before then, in the 1950s and 1960s, there were major breakthroughs in the development of novel vaccines and genetic-based diagnostics, which at the time constituted a new biological paradigm. However, the transformation of new biology into a big science that could be developed on an industrial scale, and contribute to an expectant bioeconomy, is a much more recent phenomenon and the central focus of this book.

My primary interest in new biology, and the health-related life sciences, is in understanding how it is evolving in a very specific late-twentieth- and twenty-first-century context of a highly distributed (spatially and temporally), interdisciplinary, and cross-sectoral innovation ecosystem. Within this ecosystem, multiple actors, organizations, and institutions coproduce knowledge and products and contribute to the realization of different types of value. Importantly, the social sciences have a pivotal role within this new regime of biomedical R&D, not only as observers of knowledge production processes, but as an integral part of the innovation ecosystem itself. Indeed, science and social science have coevolved in quite new and interesting ways as new biology has taken root both culturally and institutionally (Tait, 1990; Wield, 2013).

To illustrate further in the context of health innovation, in the late 1990s and early 2000s, the multinational pharmaceutical industry, which had dominated the market for conventional drug therapy for over a century with relatively simple, small-molecule compounds, was confronted with the challenge of an emerging therapeutic paradigm built around new biology (Mittra, 2008; Wield, 2013; Wield et al., 2013). At the same time, firms were struggling to sustain growth with these conventional blockbuster drugs (Mittra et al., 2012). The so-called productivity crisis within the large multinational firms (Hara, 2003; Pammolli et al., 2011) led to incremental changes in how these companies organized and managed their internal and external R&D processes. This was because the new life sciences required a much broader innovation system, involving different types of organizations and expertise, than that of traditional chemistry-based drug discovery (Hopkins et al., 2007; Mittra, 2007b, 2008; Nightingale, 2003; Rafols et al., 2014). All the major firms invested heavily in new biology and adopted similar strategies as they experimented with different ways of doing R&D. These large companies also began to strategize alternatives to blockbuster drug development, where sales in the billions of USD are required for sustained growth and the meeting of shareholder and market expectations. Personalized or stratified medicine and new biologics-based therapies, and more recently RM, emerged to challenge this conventional business model. These technologies and approaches required new and different strategies to identify and capture opportunities for value creation within a complex and evolving health innovation ecosystem. The crisis in the pharmaceutical industry, and the emergence of new biology as both an opportunity and a challenge, forms the basis of chapter 2.

However, for now I simply want to emphasize that the range of commercial and public sector actors and institutions that are now involved, perhaps by necessity more than design, in meeting the challenges and exploiting the opportunities presented by new biology, has led to the emergence of very different and often conflicting narratives of hope and promissory expectations (Borup et al., 2006; Brown et al., 2000; Bubela et al., 2012a). Translational policies and commercial strategies for exploiting life sciences have also threatened to disrupt prevailing professional and disciplinary boundaries, creating new and evolving relationships between industry, science, medicine, commerce, and society (Calvert, 2010; Cox and Webster, 2012; Martin et al., 2008; Mittra, 2013). Such changes are particularly resonant in the broader context of the emerging health bioeconomy, within which new path-breaking therapies are being developed and must find a way to successfully navigate precarious and uncertain routes to market.

Conceptualizing the Health Bioeconomy

Having defined and considered what might be unique about new biology that makes it an interesting and worthy object of study, it is important to then consider the nature of the broader bioeconomy. What is the health bioeconomy and how might it be best theorized and used to reflect on recent developments in health-related life science innovation?

The concept "bioeconomy" has been defined in a number of different ways for a number of different purposes (Schmid et al., 2012).

It began to gain traction in the late 1990s and early part of the twenty-first century in relation to economic activities and opportunities emerging from new biology. In its 2006 report, *The Bioeconomy to 2030: Designing a Policy Agenda,* the Organization for Economic Cooperation and Development (OECD) defined the bioeconomy as

> the aggregate set of economic operations in a society that use the latent value incumbent in biological products and processes to capture new growth and welfare benefits for citizens and nations. These benefits are manifest in product markets through productivity gains (agriculture, health), enhancement effects (health, nutrition) and substitution effects (environmental and industrial uses as well as energy); additional benefits derive more eco-efficient and sustainable uses of natural resources to provide goods and services to an ever growing population. (OECD, 2006: 1)

This is a broad definition of bioeconomy that focuses on new economic models and practices that must be nurtured to extract value from R&D investments in life sciences, and respond to various global challenges for both public benefit and, crucially, national competitiveness. This neoliberal flavor is also captured in President Obama's 2012 *National Bioeconomy Blueprint* (The White House, 2012) when it states: "Technological innovation is a significant driver of economic growth, and the U.S. bioeconomy represents a growing sector of this technology-fueled economy" (The White House, 2012: 1).

It is this supposed, underlying neoliberal philosophy that has been the central focus of many social science critiques of the bioeconomy (Hamilton, 2008; Parry, 2007) and the speculative value propositions and ideologies that underpin it. Cooper (2008), for instance, argues that the emergent biotechnology industries cannot be seen as separate from neoliberalism's rise as a dominant political philosophy:

> The biotech revolution...is the result of a whole series of legislative and regulatory measures designed to relocate economic production at the genetic, microbial, and cellular level, so that life becomes, literally, annexed within capitalist processes of accumulation. (Cooper, 2008: 19)

Similarly, Styhre and Sundgren (2011) describe the bioeconomy as the "economic regime of accumulation where technoscientific know-how developed in the life sciences is capable of making the lived body a principal surface of economic value creation" (Styhre and Sundgren, 2011: 3). From both these perspectives, biotechnology is coterminous with the neoliberal bioeconomy—the utility (and indeed "vitality") of life itself being determined by a particular set of dominant economic practices and subject to new and emerging processes of speculative commodification (Rajan, 2006). A number of social scientists have drawn on both Marxist and Foucauldian thought to theorize and think through the social implications of new biology and these emerging economic and institutional regimes that shape it. They have coined concepts such as "biovalue" and "biocapital" to serve as both descriptive and explanatory tools (Rajan, 2006; Rose, 2001; Waldby, 2000).

Together, these concepts are used to describe the constitutive elements of the emergent bioeconomy, which have given rise to a new "biopolitics." Some authors claim that this biopolitics goes far beyond surveillance at the population level, which Foucault described, to the constituent cells, molecules, and genomes of individuals (Helmreich, 2008). Such authors emphasize the transformative effects of life sciences and its capitalist modes of production on human bodies, tissues, identities, and sociopolitical relations. These ideas are deployed as part a powerful critique of both the increasing attribution of speculative surplus value and worth to biological material in crude economic terms, and the commodification processes that drive this phenomenon (Cooper, 2008; Novas and Rose, 2000; Parry, 2006).

Taking seriously the notion of latent value in biological processes, which the OECD emphasized in its definition of bioeconomy, there is consensus among these authors that there is something unique about the biological sciences in the twenty-first century, which justifies the promiscuous use of the prefix "bio" to conventional terms such as value, capital, and politics. It is the tight coupling of new biology with traditional capitalist modes of production and organizational processes that generate novel types of biovalue and a related biopolitics. For some, the speculative and often contradictory nature of the bioeconomy has led to an exploration of the emerging politics and sociology of hope, hype, and expectations (Borup et al., 2006; Brown, 2013; Novas 2006). I draw on aspects of this rich body of work later in the book to explore value and expectations in TM initiatives. In particular, I show how organizations in the bioeconomy display the attributes of anticipatory and promissory organizations (Pollock and Williams, 2012), which not only build expectations about the future of medicine, but also actively shape technological options and therapeutic value chains.

This particular approach to life sciences and the bioeconomy has not gone uncriticized. For example, Birch and Tyfield (2012) have critiqued many science and technology studies (STS) approaches for both fetishizing the biological sciences, and conceptualizing an ever growing number of ambiguous bioconcepts that lack any meaningful explanatory power. Furthermore, they suggest that STS theorists have, due to their focus on technoscientific features of the bioeconomy, given too little attention to the transformations of the underlying economic and financial process of contemporary capitalism (Birch and Tyfield, 2012: 301). The essence of Birch and Tyfield's critique is captured in the following statement:

> We highlight the problematic adoption of Marxist language in these bio-concepts without the necessary adoption of Marx's theoretical formulation of the labor theory of value (LTV) underpinning key terms like value, capital, and surplus value. In adopting Marxian concepts, the fetishization of the "bio" has meant that – to different degrees – STS scholars like Waldby, Rose, Rajan, and Cooper have missed an opportunity to update the understanding of the bioeconomy in light of the financial and economic restructuring of the economy. (Birch and Tyfield, 2012: 301)

The authors go on to argue that it is important to consider "asset-based" economic processes as an integral part of the bioeconomy, rather than the "commodity-based" processes that STS scholars tend to prioritize and which are captured in, for instance, the rhetoric of "life as surplus" (Cooper, 2008). The former, according to Birch and Tyfield, is a tangible or intangible resource that both produces value and entails value as property, while the latter is simply an object produced for exchange. Furthermore, terms like "latent value" and "surplus value" imply that there already exists intrinsic, but yet untapped, value and vitality in biological material and processes. This neglects the fact that such things accrue value over time through the immaterial labor of commercial and public institutions.

Birch and Tyfield's work, captured in the quotations above, not only subjects the narrow, commodity-based approach to critical analysis, and questions the newness of concepts such as biovalue and biocapital, but does so in the context of what the authors consider to be real changes in economic processes that shape the bioeconomy. In particular, they identify a "...shift in value creation from productive to immaterial labor...a financialized-rentier regime of accumulation; and...the shift from commodity-based to an asset-based market exchange" (Birch and Tyfield, 2012: 301) as three key transformative processes underpinning the neoliberal bioeconomy.

Birch (2006) has also suggested that policy discourses centered on national competitiveness, and the economic representations and practices underpinning the bioeconomy, have naturalized and served to justify policies, institutions, and governance regimes that have shaped the current innovation system for biotechnology. Hilgartner (2007) shares this sentiment when he suggests that organizations such as the OECD are ambitious "anticipatory enterprises" that both predict and shape the emergent bioeconomy, such that the economic representations serve an important performative function. Birch (2007) has also described the bioeconomy as a "virtual abstraction" of economic practices in which benefit and potential "...are intertwined concepts...repeated numerous times throughout this policy literature, which essentialises and naturalises the claims made about its innovative potential" (Birch, 2007: 89). Other authors have been even more explicit in describing the bioeconomy as a political project and promissory construct to support neoliberal capitalism, rather than a strictly scientific, technological, or economic endeavor (Goven and Pavone, 2015).

I do not wish to discuss in-depth the subtle nuances and minutiae of these particular debates about the nature of the bioeconomy. I simply want to emphasize that the policy assumptions and political traction that has driven a global bioeconomy agenda in the twenty-first century have been questioned by a number of social scientists, and particularly STS scholars. It has also encouraged a range of new approaches for studying the evolution of life science industries and the socioeconomic values underpinning them. The approaches so far outlined highlight the gap between speculative and actual economic value. They also raise serious questions about transformative expectations and the performativity of market-based policies and understandings, particularly their impact on science and society. For some, the negative implications of the current regime tend to be emphasized or implied, but it is not always clear where this should lead in terms of normative policy change. Much of this work has provided valuable conceptual and theoretical tools for understanding various components of the bioeconomy, and the foundations on which it has been constructed. However, we must go further to better understand the dynamics of the innovation ecosystem and its different stakeholder interactions and expectations of value.

It is my aim in this book to consider more pragmatically what the key challenges are to the success of a vibrant health bioeconomy, notwithstanding the validity of some of the critiques concerning its purported neoliberal assumptions and modus operandi. I also want to think through how the challenges and limitations to exploiting new biology for both therapeutic and economic benefit might be overcome. In so doing, I consider if new conceptual approaches to value and valuation practices in health innovation might enable a more sophisticated analysis of the innovation ecosystem for the health-related life sciences and its nascent bioeconomy.

Value in the Health Bioeconomy

There are a number of challenges facing the health bioeconomy and the transformation of life science knowledge and expertise into a viable biobusiness. Gary Pisano's 2006 book, *Science Business: The Promise, the Reality, and the Future of Biotech*, was an important and at the time well-received account of some of the structural problems and challenges facing the life science sector, which were preventing it from making the anticipated contributions to the bioeconomy and public health. Pisano's great insight was in revealing some of the reasons why the biotechnology sector was not thriving and successfully turning what was believed to be revolutionary science into commercially viable and high-value products. On the one hand, Pisano highlighted some of the distinctive features of the biotechnology industry, particularly the high risk and uncertainty of R&D. Drug innovation is expensive, time consuming, and highly risky. It can take up to 20 years to take a new therapy from discovery to the clinic, and at various stages along that development pathway the therapy is liable to fail. This is an industry unlike any other. On the other hand, and perhaps more interestingly, Pisano pointed to the fundamental clash of values, norms, and practices between the worlds of science and business. The challenges of funding R&D and successfully integrating different knowledges and practices were highlighted by Pisano as key reasons for the gap between the promise and the reality of contemporary biotechnology.

However, Pisano's analysis was quite narrowly focused on intellectual property regimes (particularly patents) and finance/funding models for the biotechnology sector. It lacked a more systemic understanding of the broader innovation ecosystem and the full range of enablers and constraints that shape its evolution. Although intellectual property regimes constitute an important element in the innovation process, and play a strong role in shaping the innovation ecosystem and the locus of value within it, they are not the central focus of this book. Much has been written about how patents in biotechnology enable or restrict innovation, determine the value of new therapies and the viability of different business models, and in a sense govern the innovation process (Gold et al., 2007). I do not wish to downplay their importance, but I contend that they are just one part of the innovation story. I discuss patents and intellectual property only insofar as they help inform and illustrate particular case examples (such as business models and reimbursement systems for stratified medicine or RM). In this book, I want to take some of Pisano's key insights, particularly in the context of how different values and norms emerge in health innovation, to a different level by exploring the innovation ecosystem in a much broader sense. This requires consideration of how regulations and policies, in the context of changing social and clinical expectations and constraints, shape innovation and R&D practices and create new organizational principles. In this context, we need a way of better conceptualizing the role of value and values, which will be a prevailing theme throughout the book.

All the different accounts of the bioeconomy so far discussed rest on a particular understanding of value and/or values. The two concepts have, for historical reasons, been treated as separate within the social sciences. Stark (2009) traces this demarcation to what he calls "Parsons Pact," when the American sociologist Talcott Parsons attempted to delineate the boundaries of sociology so as to placate the economists who felt their territory was under threat. Parsons suggested that sociology would study *values* while economists would be left alone to study *value*. The economic and social facets of "value" were now decoupled and have largely been treated as distinct domains ever since. Put simply, and perhaps crudely, sociologists study the subjective social relations underpinning the economy, while economists, and the calculative sciences, study "objective" economic value, reflected, for instance, in the market price. However, there is now a growing network of social scientists trying to rejoin value and values to develop more pragmatic studies of "valuation as a social practice," which tries to capture both the objective and subjective elements of value (Beckert and Aspers, 2011; Dussuage et al., 2015; Helgesson and Muniesa, 2013; Lamont, 2012). Indeed, the very definition of value has always had both an economic and noneconomic component, but in common parlance value does tend to evoke the former rather than the latter. Furthermore, this emerging body of work emphasizes that there is no real intrinsic value to any object (value is not an attribute), nor transcendental values that exist outside of social norms and practices. Some go further by arguing that value emerges from firms and entrepreneurs' ability, through business models (which are essentially constructed narratives of how things can be made to work), to build networks and system structures that will permit the realization of value (Perkmann and Spicer, 2010). These perspectives enable us to think about value in a much broader, interdisciplinary sense, rooted in the complexity of social practices and institutional assemblages where valuation and worth are constantly negotiated and established through a variety of tools and technologies (Helgesson and Kjellberg, 2013).

In this book, I use the terms value and values to capture both the economic and noneconomic processes of evaluation and different accounts of benefit or worth in the context of the health innovation ecosystem and its bioeconomy. This, I argue, allows us to avoid over-emphasizing the pecuniary aspects of value, and reducing innovation systems and the bioeconomy in health to the crudest of economic metrics. As Beckert and Aspers rightly note, value and valuation are matters of concern even in the absence of money (Beckers and Aspers, 2011: 3). Fourcade (2011a) suggests that money actually tends to conceal the "real" essence of things because it simply conflates economic value with market price. Indeed, some argue that there are different "regimes of value" (Appadurai, 1986), but no single scale under which different types of value may be subsumed (Beckers and Aspers, 2011: 6), despite the common tendency to reduce value simply to matters of finance. So value and price are far from synonymous, and nowhere is this clearer than in the case of pharmaceuticals, as we shall see later in the book.

Helgesson and Muniesa nicely capture the multiplicity of value when they write:

> What things are worth can be manifold and change—and these values can be conflicting or not, overlapping or not, combine with each other, contradict each other. All, or almost all, depends on the situation of valuation, its purpose, and its means. Broad segmentations such as the distinction between "economic" and "non-economic" value can make sense at some level, only the devil is in the detail. Something valued as a financial asset, for example, can be valued differently by different accountants or different investors. And then this thing can be valued in an entirely different way in other circumstances (i.e., not as financial asset, but as a political project, as personal property, you name it). (Helgesson and Muniesa, 2013: 7)

This broad and more inclusive approach to value and valuation, which encourages us to unpack the different ways in which value is enacted or performed in specific professional domains and social contexts, opens up new avenues for studying things like the bioeconomy and the organizational practices of new biology and clinical medicine. Stark (2009) argues that the plurality of principles of evaluation that operate within society suggests that any social order (such as a modern economy) contains multiple "orders of worth," which determine value and form the very basis of calculation and rationality (Stark, 2009: 11). Kelly and Geissler (2011) have applied this concept to the realm of modern clinical trials, describing them as generative of many different orders of value and intersecting the worlds of both commodities and public goods, which problematizes the conventional distinction between fiscal and moral virtues (Kelly and Geissler, 2011: 3). Mol's (2003) study of the multiplicity of meanings, practices, and, I would suggest, the different orders of worth within a hospital ward for the treatment of atherosclerosis is also within the spirit of this plural approach to value and valuation practices, as is Fourcade's (2011b) analysis of the different ways in which economic value is ascribed to intangible things such as "nature" when calculating liabilities from oil disasters. In the latter example, Fourcade revealed how political and cultural specificities led France and the United States to develop very different methods, metrics, and evaluative criteria to determine the appropriate fiscal penalties associated with the destruction of valuable biomass.

Espeland and Stevens (1998) have looked at the related problem posed by the need for "commensuration," which is the ability to compare different entities by a common metric. This is particularly pertinent to many innovation ecosystems in health R&D where there are multiple, overlapping value chains and systems, including public and commercial organizations subject to very different metrics of evaluation and criteria for success. Commensuration is also relevant to the "abstract value" of commodities, which is nicely captured in Brown's (2013) study of use and exchange value in the cord blood economy. In this case, it is not always useful to strictly delineate "use value" and "exchange value" on the grounds that the former provides public incentives for the bioeconomy (contribution to public health and a broader moral order of worth) and the latter provides incentives for commercial organizations (producing objects with proprietary rights for exchange in markets).

However, there is a danger in conceptualizing this multilayered approach to value and valuation practices as simply a conflation of the economic and the ethical/moral, where the latter and perhaps more subjective judgment of moral value and worth is given undue priority. In the social sciences, this can easily manifest as a mechanism to permit values-based arguments (in the transcendental sense) against various technologies to be prioritized. Taken to an extreme, it can ultimately threaten at a very early stage of development, under the auspice of upstream engagement and more recently responsible research and innovation (RRI), potentially beneficial innovations (Tait, 2009). It can also lead to the marginalization of commercialization processes and economic value as an important facet of innovation ecosystems and therapeutic product pathways, which is damaging if ethical and moral judgments, and the often overused trope of "societal concern," take precedence and are subsequently valorized in academic and policy discourse. In this book, my interest in adopting a broader and more inclusive definition of value (which takes seriously the notion that economic value emerges from multiple types of other values and valuation practices) is to allow for a more sophisticated description and analysis of how health innovation systems are made to work in practice. I want to unpack how different R&D options, knowledge, and expertise; and the more tangible benefits of innovation are valued (or not) by different stakeholder communities who must nevertheless work together to develop new therapies. In a sense, actors and institutions in the contemporary health-related biosciences, each with their own expectations of value and benefit, must "muddle through," to borrow a concept from Lindblom's (1959) seminal essay on scientific administration and decision making.

So value and values, in the context of how I use the terms throughout this book, do not relate strictly to pecuniary matters on the one hand, and intrinsic ideologies and belief systems on the other. The latter would simply reify the notion of transcendental values, which I think is important to avoid in this context. Nevertheless, I do want to capture the nuances and differences of multivalent institutional values and valuation practices (scientific, clinical, commercial, and political), which determine how objects (in this case novel technologies, therapeutic products, and the processes that produce them) are valued on the basis of both economic and noneconomic criteria. Even if crude economic value (established by market price) is the ultimate driver of some aspects of the health bioeconomy, and the basis on which it is deemed a success or failure, this broader approach to value and its underlying practices is required to fully understand the nature of health innovation ecosystems and how they can deliver the benefits promised to various stakeholders. Crucially, the relevant stakeholders now include patients and publics. For example, Mazanderani et al. (2013) highlight the important role of patients in producing "biographical value" through "illness narratives." They talk about the commodification of illness experiences in terms of the "rise of different and overlapping markets in which illness narratives are produced, circulated, used and exchanged, generating value in different ways for different people" (Mazanderani et al., 2013: 891). These narratives, which may be published through various social media, have value for other patients, health charities, and patient groups; as well as those who may want to measure the quality of health care. This is a nice example of how value should be seen in its broadest sense, encompassing both economic and noneconomic elements and implications. I discuss this in more detail in chapter 6.

This idea of broadening our idea of value is being debated not only by social scientists, but also by the science, industry, and policy communities responsible for advancing new biology. A series of articles published in the Lancet in 2014, for example, came with the tagline "Increasing Value, Reducing Waste." Each article considered how the value of medical research could be increased by reducing known sources of waste. Examples included a more robust and transparent process for setting research priorities (Chalmers et al., 2014), increasing access to all research to minimize bias and improve data sharing (Chan et al., 2014), and reducing burdens of regulation and management (Salman et al., 1014). Value in this context refers to not only conventional economic value that drives the bioeconomy, but also value in terms of specific patient and societal benefit. It also encompasses benefit to science and clinical practice more generally, particularly in terms of improving the quality of research and health care and achieving greater efficiency with existing resources.

However, Cutler and McClellan (2001) caution that while reduction of waste might be considered a valuable endeavor, it must be balanced against the potential for less rapid innovation. Indeed, I would argue that some waste is a natural part of any ecosystem, even an innovation one (there is no real waste in a natural ecosystem). Furthermore, who ultimately defines what is wasteful? Much of the recent interest in capitalizing on "big data" initiatives and opportunities (ABPI, 2013; Kayyali et al., 2013) is centered on this supposed need to minimize waste and increase the efficiency and value of health research. However, as I explain later in the book, the wealth of new information made available by advancements in new biology may problematize our notions of waste and efficiency. Furthermore, efficiency drives rarely prove to be a panacea for innovation challenges, as we shall see in chapter 2 in the context of multinational pharmaceutical companies and their investment in life science technologies for drug discovery.

Nevertheless, questions are now being asked by policymakers, industrialists, scientists, health-care providers, patients, and taxpayers about the value and worth of medical research and life science innovation. Such questions operate with an ever more broad and sophisticated concept of value. For example, recent developments in "value-based pricing" for medicines, which I discuss in chapter 5, entail a much broader notion of long-term benefit and patient value, specific to particular therapy areas, than conventional reimbursement systems (BMJ, 2013). Narayan et al. (2013) claim that

> in major depression, value may be defined as the ability to rapidly resume social and work responsibilities; for pain, it may be defined as the ability to quickly resume physical activities of choice; and for Alzheimer's disease, it may be defined in terms of benefits that allow patients to remain independent for longer...in such an environment, the main driver of improved outcomes and meaningful benefits may not be innovative therapeutics alone but an ecosystem comprising the therapeutic and wrap-around tools and services. (Narayan et al., 2013: 85, 86)

From this perspective, integrated solutions for health care are needed. This compels us to recalibrate our notional ideas of value as high-tech life science approaches must coevolve with other organizational innovations in the broader ecosystem to deliver longer-term value in terms of both patient benefit and sustainable commercial revenue. The authors proceed to argue that there are continuing challenges for these new integrative solutions in health care, particularly in terms of regulatory pathways and business models. Porter (2010), looking at the issue from a US perspective, points to the organizational and information systems of health care that make it difficult to measure and provide value:

> Providers tend to measure only what they can directly control in a particular intervention and what is easily measured, rather than what matters for outcomes. For example, current measures cover a single department (too narrow to be relevant to patients) or outcomes for a whole hospital, such as infection rates (too broad to be relevant to patients). Or they measure what is billed, even though current reimbursement practices are misaligned with value. Similarly, costs are measured for departments or billing units rather than for the full care cycle over which value is determined. (Porter, 2010: 2478)

This is a continuing challenge for the successful management of health-care systems and the sustainability and resilience of a health bioeconomy. My colleagues and I at the Innogen Institute have responded to these challenges in recent years by developing a unique

approach to innovation ecosystem analysis for new and complex therapies that challenge existing product development pathways and value
chains (Mastroeni et al., 2012; Mittra and Tait, 2012; Mittra et al.,
2015). Case examples and relevant vignettes from this body of work is
illustrated later in the book. But these challenges of determining the
value and benefit of new life science therapies, which depend largely
on the level of unit analysis, the calculative practices used, and the
discrete parts of the innovation system that are ascribed value and
considered to be "of worth," can be better unpacked and understood
within this broader definition of value.

Another illustrative example comes from the Dutch innovation
context, where the concept of "valorization" has been used to denote
this conjunction of the economic and noneconomic aspects of value.
Stemerding and Nahuis (2014) talk about the challenge of the "valorization of knowledge" and describe how valorization was used by
Dutch policymakers toward the end of the 1990s to define and direct
"...the process to create value from knowledge by making it available
for economic and/or societal use and by translating it into competitive products, services and new business" (Stemerding and Nahuis,
2014: 80). This approach has become institutionalized within the
Dutch innovation system and has led, according to supporters of
valorization, to the use of indicators and evaluative practices that go
beyond crude, instrumental economic metrics as the basis for value.
The valorization concept is particularly pertinent to collaborative
public-private partnerships, which I discuss in detail in chapter 4.

So, the conflation of economic and noneconomic value should not
necessarily lead to an anything goes approach to the study and critique of innovation ecosystems and/or a prioritization of the more
difficult to quantify ethical and subjective social values. However,
understanding how different actors involved in innovation processes
generate, negotiate, and integrate different notions of value and
orders of worth to make R&D work in practice is an important and
much needed contribution social science can make to the biomedical
innovation process. Frow (2008) talks about her interest, in the context of synthetic biology, in understanding what practitioners count
as worth knowing and to what ultimate end. She has explored how
the field is being shaped by different valuation practices. In this book,
I am interested in exploring this in the context of a broad range of
active therapeutic R&D processes for novel therapies based on new
biology that is driving, but also challenging, specific enactments of
the health bioeconomy.

Overarching Aim of the Book

My broad aim in this book is to problematize those accounts that simply dismiss new biology as hype, or see its impact on organizational norms and practices as marginal, or perhaps even negative. I also want to explore the transformative effects of new biology and the growing bioeconomy on the structure and organization of R&D within health innovation ecosystems, rather than the more vague and ambiguous realm of biopolitics and social relations. I do not begin with a strong normative view about the neoliberal aspects of the bioeconomy. Neoliberalism, which is quite a vague and ill-defined term, is often used pejoratively when talking about emergent biotechnologies and the institutional and organizational relationships that nurture them. Indeed, I suggest that the very nature of new biology, and the diverse and complex value chains underpinning its different application areas, means that it is perhaps inevitable that a broadly neoliberal framework has so far guided its development and continues to shape its R&D pathways. The key is to better understand these new R&D processes and practices, and the multiple values and valuation practices that are being performed by different actors within the health innovation ecosystem.

In building the argument, I reflect on those insightful accounts and conceptual approaches that have explored how different promissory visions and expectations in science and technology shape R&D policy and practices (Borup et al., 2006), and engender new communities of promise and organizational principles in translational life sciences (Martin et al., 2008). I draw on work that has discussed how diverse and often contested notions of value in the bioeconomy drive science and technology options, and often create new "bio-objects" that challenge conventional boundaries and the status quo of public and commercial research (Webster, 2013). I argue that the perception that there is a problem of "translation" in health R&D, and the subsequent changes in policy, funding, and strategic behavior of commercial and public sector innovators, has had a material effect on the "doing" of R&D and the way it is valued by different professionals and experts that are embedded in the innovation ecosystem. In this context, the concept of novel experimentation applies not only to the basic research underpinning new biology, but also to the organizational structures, management strategies, and policies that are being implemented to capitalize on new biology. Just like basic research, there are successes and failures in these organizational and policy experiments. This notion of "trial and error" is an inevitable feature that drives progress, as is the building of future expectations and visions. If we accept this notion, we need not fall into the trap of judging any policy or organizational experiment that does not meet its initial objectives or expectations as a failure, or as having no value in the broader context of the evolving innovation ecosystem.

A key argument I make in this book is that the emergence of TM, as both a general philosophy and set of specific industry and policy-driven initiatives to ensure novel therapeutic products make it to the clinic, must be seen in the much broader context of the systemic challenges facing multinational commercial drug development, the changing relationship between basic and clinical research, the emergence of new organizational relationships and interdisciplinary ways of "doing" R&D, and regulatory/policy challenges of assessing risk and benefit of new types of therapies. The concept of a health bioeconomy is central to this, as it represents a vast range of economic and noneconomic activities and valuation practices. As Wield (2013) usefully points out, health brings together two important but often separate aspects of the bioeconomy: innovation in new therapies (with very long lead times) and health policy and services, which are driven by issues of treatment cost and access to health systems. There is often little integration of the policy and innovation perspectives, so it is important to provide a more systemic and integrated account.

Empirical Data and Case Studies

The data informing the arguments made in this book emerged from a range of projects I have conducted either solo, or in collaboration with colleagues within the Innogen Institute at the University of Edinburgh. Together, the complete dataset includes 15 interviews I conducted with senior R&D managers and scientists within big multinational pharmaceutical firms and small and medium-sized biotechnology companies in Europe and the United States (conducted in 2004 and 2005); 35 interviews and a workshop I organized with key practitioners involved in various translational initiatives, including senior academic scientists, clinicians, health-care service managers, policymakers, and regulators (conducted between 2010–2013); 15 interviews with investors, scientists, and academics in the field of RM (conducted with colleagues in 2013); data from three workshops and a small number of interviews on product development strategies for an ESRC/TSB RM project (2010–2011); and field notes from attendance at various industry and stakeholder conferences and workshops over the past 12 years. Most of the data derives from qualitative research interviews (which are fully anonymized throughout) and workshop data, but also includes significant secondary gray literature and policy document analysis.

Throughout the book, rich case study examples are used to tease out key themes, such as the organizational restructuring of the pharmaceutical industry and its attempt to integrate new biology, challenges and opportunities of personalized and stratified medicine, and the associated development of new diagnostic biomarkers and related devices to better target therapies to specific patient subpopulations, and RM, which promises more fundamental changes to the nature of therapy and translation to the clinic. My aim is to integrate microlevel critical analyses of institutional and organizational norms, disciplinary practices, and relationships in emerging technology areas that challenge the status quo, with more systemic macrolevel analysis and evaluation of how regulation, policy, and markets shape technology pathways and options within complex innovation ecosystems. Further information about specific datasets used in this book and reference to projects within which data were collected are described in more detail in the notes within individual chapters.

Outline and Structure of the Book

The book comprises seven chapters. In the following chapter, I address the question: *What impact does the life sciences have on the organizational structure, commercial strategies, and R&D practices of the pharmaceutical industry?* Looking back to the late 1990s and early 2000s, I draw on interview data to critically explore the strategies developed within the largest multinational pharmaceutical companies as they tried to respond to a so-called productivity crisis, and exploit the emerging opportunities presented by new molecular biology. After providing a brief history of pharmaceutical innovation, I reveal how companies began to experiment with new ways of organizing and managing R&D, strategically using mergers and acquisitions, and alliances with smaller biotechnology companies, to appropriate external knowledge, skills, and expertise and build new capabilities in life sciences. I also address the pharmaceutical industry's struggle to identify new sources of value for life science-based therapies, and think through potential future routes to market—a market begrudgingly accepted by large companies to be unlikely built on blockbuster small-molecule drugs.

In chapter 3, I address the notion of a "broken middle" in the health innovation pathway, which led to the emergence and prioritization of translational medicine/research as a powerful commercial and public policy strategy embraced by industry, government, and the scientific and clinical communities. The key question to be addressed in this chapter is: *What perceived challenges, opportunities, and practitioner values in health innovation have driven a new translational policy agenda, and with what consequences for the bioeconomy?* In this chapter, I explore diverse practitioner definitions and multiple meanings ascribed to translation, as well as the nature and underlying realities of the problem it seeks to address. I then think through some of the practical implications and effects this has had on health policy and innovation, and indeed the very relationship between the public and commercial sectors in health research. A particular focus in this chapter is the strategies of organizations such as the National Institutes of Health (NIH) in the United States, and major funding bodies in Europe and the United Kingdom (Technology Strategy Board and Medical Research Council), to capitalize on new biology and generate long-term scientific, commercial, and clinical value and, ultimately, patient benefit.

In chapter 4, I move from a broader macrolevel analysis to look at changes in R&D practices. The question to be addressed in this chapter is: *In what ways has the "doing" of R&D been reshaped by the institutional and organizational restructuring precipitated by translational policies and how are stakeholder expectations and values recognized and managed?* In this chapter, I address the crucial organizational and institutional impact of the translational policy agenda. I do this in terms of how interdisciplinary and cross-sectoral collaborations are having a material effect on what it means to do R&D in the laboratory and the clinic, and are determining what is ultimately valued. If interdisciplinarity disrupts conventional professional and sectoral boundaries, it may also reveal tensions around different notions of value and benefit. Translational R&D structured around new biology may bring many new opportunities for therapy development, but it must also deal with institutional constraints and manage different expectations about scientific, clinical, economic, and social values and benefit. In this chapter, I critically explore some of the salient organizational and institutional changes that are being precipitated by this public and commercial interest in new ways of doing life sciences. Three rich case studies of public-private partnerships are used to illustrate the key arguments, alongside interview data from key practitioners.

Chapter 5 is driven by the question: *How has new biology both challenged and transformed conventional regulatory systems and the resilience and adaptive capabilities of health-care systems to innovative therapies?* The central focus here is on the broader regulatory and policy challenges of developing radically new approaches to therapy that do not have established routes to market and conventional business models, value chains, and regulatory precedents to take them from the laboratory to the clinic. Many of these technologies, if they are to be successful, must find a way to fit into, or transform, existing health-care pathways and navigate complex and often sclerotic regulatory and reimbursement systems. In this chapter, I reflect on some of the key regulatory developments and strategies in the United States and Europe before exploring path-breaking technologies and approaches (using the case examples of RM and stratified medicine) that challenge our conventional regulatory systems. I also explore the implications of new therapies for health technology assessment and reimbursement systems, particularly in the context of new approaches to value-based pricing. This chapter deals with the problem of how new technology can break through existing institutional logics and rationalities built into an already well-established system of institutions and organizations that may have a vested interest in maintaining the status quo. I also highlight the direct impact that regulation and policy can have on the trajectory of new technology and therapeutic innovations, and the ability of researchers, particularly in the public sector, to deliver viable health solutions to the clinic.

In chapter 6, I address the question: *What are the implications of the changing role of patients and publics in the new health bioeconomy, and how can their expectations and values be better understood and managed?* Here, my focus shifts to the increasing role patients and publics are expected to play in health innovation and research. I consider the broader and long-term implications of this for the development of new therapies. The success of new biology and the health bioeconomy requires a far greater and more active role for the public and patients in health and clinical research, in addition to the collaboration of diverse institutional actors with very different notions of the value of the science for commerce, medicine, and society. These issues are becoming increasingly salient and powerful as patients actively lobby for better access to innovative therapies, and policymakers strive to increase value and minimize "waste" to meet these growing expectations and societal needs. The issue of "big data," and how this may be used to improve innovation, is a crucial factor to take account of in the context of the health innovation ecosystem. This chapter tries to capture and critique the different ways in which the patient and broader publics have become valued participants in the R&D process.

In the final concluding chapter, the key themes addressed in the substantive chapters are summarized and I return to the core notion of value in the emergent health bioeconomy. The question to be addressed in this chapter is: *What is the future for therapy in light of the many experiments in translational medicine; the nature of the evolving bioeconomy and the constellation of value therein?* Here, I consider the long-term consequences—for industry, science, medicine, and broader society—of the current and largely experimental policy initiatives and strategies that are arguably precipitating change in the nature and organization of R&D, and the various interdisciplinary and cross-sectoral practices that have been engendered.

Chapter 2

Crisis in the Pharmaceutical Industry and the Promise of New Biology

Introduction

When Tadataka Yamada arrived at GlaxoSmithKline (GSK) as the new CEO, following the merger of SmithKline Beecham and Glaxo Wellcome in 1999, he found the company in serious decline in terms of its productivity and innovative capabilities. His first major decision was to initiate a new creative approach to research and development (R&D); restructuring the organization and management of R&D by dividing the company's 1,900 discovery scientists into six Centres of Excellence in Drug Discovery (CEDDs), each focused on a specific therapeutic area. A seventh CEDD focused on biopharmaceuticals and a "virtual" Centre of Excellence for External Drug Discovery were later established. The impetus for Yamada's restructuring program was his growing skepticism about the traditional pharmaceutical model of R&D that the large multinational firms had pioneered. In particular, he questioned its reliance on a very centralized and bureaucratic decision-making structure, and was uncomfortable with the artificial distinction between the discovery and development phases of research. The philosophy of the CEDD "hub and spoke" model was to confront GSK's productivity problem by creating relatively autonomous and geographically diverse R&D units that would replace many of the functions of centralized management systems for the middle stage of R&D (clinical development), where many drugs tend to fail.

GSK's major restructuring initiative, which is described in more detail by Mittra (2008) and Huckman and Strick (2005), initiated a broader trend in the sector, as large firms tried to capture the innovative spirit of the smaller biotechnology companies that were driving innovation in new biology. In this chapter, I reflect on the history of pharmaceutical innovation and illustrate how industry adapted to the emergence of disruptive life science technologies and the broader challenge of how to derive value from a therapeutic paradigm rooted in new biology. The overarching question is: *What has been the impact of life sciences on the organizational structure, commercial strategies, and R&D practices of the pharmaceutical industry?*

Over the past 30 years, therapeutic innovation has evolved such that new biology and life science technologies have become a major part of R&D. For early stage drug development, automated high-throughput molecular screening technologies and systems biology have been used to identify new drug compounds and their biological targets. More recently, therapies based on advances in new biology have emerged, including recombinant therapeutic proteins, monoclonal antibodies, and regenerative medicine (RM). These are step-change innovations with different product development pathways, regulatory implications, and value chains from the more conventional small-molecule drugs. The foundations of the health bioeconomy, and the realization of different types of value within industries such as pharmaceuticals, depend on the continued success of this transition from familiar old chemistry to a more complex and idiosyncratic new biology.

The drivers for change in how drugs are discovered and developed were partly the adoption of new strategies by large multinational pharmaceutical companies, as traditional small-molecule drug development matured in the early 1990s (Mittra et al., 2011) and the "blockbuster" business model came under considerable strain.[1] However, a transformation in the innovation ecosystem for new therapies was also driven by advances in life science technologies and the emergence of small, innovative biotechnology firms that disrupted the status quo and established value chains. These companies brought hope and expectation of a revolutionary therapeutic paradigm built on the biological sciences. Although the reality was incremental evolution rather than biotechnology revolution, as I discuss later, the shift toward an R&D platform built on the life sciences was challenging from an organizational and management perspective. Large pharmaceutical firms, such as GSK, underwent a prolonged period of experimentation in how they organized and managed their R&D pipelines and made investment decisions. This shaped broader innovation system dynamics and practices.

In this chapter, I unravel the complex story of how new biology and life science technologies shaped the organizational structure, commercial strategies, and R&D practices of a beleaguered pharmaceutical industry at the end of the twentieth century. This was an industry with a long and prosperous history that struggled to adapt to the interdisciplinary life sciences and new knowledge and technology dynamics that they brought. Drawing on data from interviews conducted in 2004/2005 with senior representatives from the pharmaceutical industry,[2] I reveal the challenges that beset the companies in the mid-1990s and explore how they organized and balanced internal and external R&D—using mergers, acquisitions, and alliances to appropriate knowledge and products from external innovators and build capabilities in life sciences. However, firms were faced with the challenge of how to identify and extract value from therapies based on new biology and create sustainable routes to new markets. They came to accept, albeit reluctantly, that new markets would unlikely be built on the blockbuster small-molecule drugs that had served them so well for decades. Conventional drug pipelines had become seriously sclerotic by the late 1990s, and this created a lingering sense of crisis within the sector.

A Brief History of Pharmaceutical Innovation

Throughout its long and mostly prosperous history, the pharmaceutical industry has adapted to continual scientific and technological change. The evolution of pharmaceuticals has been driven by both incremental and disruptive innovations that have shaped commercial strategy, markets, and institutional and organizational norms and practices. According to Malerba and Orsenigo (2002), the history of industrial pharmaceuticals can be usefully cleaved into three epochs, and I follow their broad categorizations in presenting the historical background to modern drug development.

The First Epoch: 1845–1945

The first epoch, which covers the period 1845–1945, constituted very little of what we might now recognize as innovative drug development, particularly in the early part of this period. It was based on what now would be considered crude scientific methods. Until the end of the nineteenth century, effective therapeutic medicine was virtually nonexistent. Rang (2005) illustrates this point by referring to the first addition of the *British Pharmacopoeia*, which was published in 1864. He notes that of 311 preparations that were listed, 187 were derived from plants (although only nine had been purified) and the rest were mainly inorganic chemicals and a few animal products. Few of these substances had what would now be considered therapeutically relevant components (Rang, 2005: 4).

The prevailing business model in the mid- to late nineteenth century was the marketing of "natural" products with unpredictable and often dangerous effects. Since chemical companies lacked the capability to purify these products and remove contaminants through standardized manufacturing processes, there was inherent uncertainty about product efficacy and safety. A more advanced approach only began to take shape in the second half of the nineteenth century, with the emergence of the first generation of therapeutics derived from the purification of natural products and the isolation of their active ingredients (Nightingale and Mahdi, 2006). Nightingale and Mahdi argue that greater standardization, enabled by new purification processes, provided much more certainty about clinical use. Hopkins et al. (2007) describe this early period of pharmaceutical innovation as based on what they call an "extractive heuristic." What this means is that companies simply isolated natural medicinal compounds from exotic plants that were provided in abundance by the many botanical expeditions at the time. Stronger patent protection for chemicals was enacted to incentivize innovation, and capabilities in synthetic chemistry were further developed and used to improve the performance of these natural alkaloids. The result was the development of many new antipyretic drugs (for the treatment of fever) at the end of the century, with aspirin being the most notable example (Hopkins et al., 2007: 568).

Major milestones in the birth of a nascent industrial sector for chemical therapies were the elaboration of cell theory by the German pathologist Rudolf Virchow in 1858, the birth of pharmacology as a scientific discipline through the work of Paul Ehrlich, and the widespread acceptance of Pasteur's germ theory of disease (Rang, 2005). However, in order for this fledgling chemical industry to deliver effective therapies for the amelioration of disease, it had to do two things. First, it had to identify and better understand the underlying physiological mechanisms of disease, which required advances in the field of biomedicine. Second, it had to embrace advances in chemistry to improve knowledge of the basic structure of active molecules and develop techniques to modify them. This was necessary to reduce adverse side effects and enhance efficacy. However, Reiss and Hinze (2000) describe the early period of synthetic organic chemistry, which produced numerous small molecules that could be tested for biological activity, as largely based on trial and error and deductive reasoning. Nightingale and Mahdi argue that without profound knowledge and understanding of the relationship between chemical structures and biological activity—or the emergence of a "biological heuristic" to complement the "extractive heuristic" (Hopkins et al. 2007)—innovation was unguided, costly, and time-consuming.

Nevertheless, these early advances began to shape a new industry dynamics as various synthetic drugs began to be manufactured and tested from the 1880s onwards by new chemical companies that had combined pharmaceutical and synthetic dye and textile divisions. The knowledge diffusion from the textile and dye industries to pharmaceuticals is an important and interesting development in the history of synthetic organic chemistry. As Rang (2005) notes, Paul Ehrlich, who some would describe as the founder of molecular pharmacology,

> became interested in histological stains and tested a wide range of synthetic dyes that were being produced at that time. He invented "vital staining"—staining by dyes injected into living animals—and described how the chemical properties of the dye, particularly their acidity and lipid solubility, influenced the distribution of dye to particular tissues and cellular structures. (Rang, 2005: 5)

It was only a small innovative step to move from using dyes to stain cells, to actually testing them for pharmacological activity. Garavaglia et al. (2006) point out that the first entrants to this new chemical industry were Swiss and German companies, such as Bayer, Hoechst, Ciba, and Sandoz, which built on their knowledge and capabilities in organic chemicals and dyes and began to apply this to the therapeutic domain. They were later joined by British and American manufacturers such as Eli Lilly, Pfizer, Warner-Lambert, Burroughs-Wellcome, and Wyeth (Garavaglia et al., 2006: 237).

In the first few decades of the twentieth century, drug discovery driven by synthetic chemistry became the established pharmaceutical model, with biology struggling to keep pace with the rapid technological advances. Nevertheless, an incipient large-scale and centralized R&D process was beginning to be normalized, and the foundations of a multinational chemical industry established. Key innovations in the second and third decades of the twentieth century included the extraction and development of synthetic analogues of various steroids, including estrogen, and hormones such as insulin. Eventually, antibiotics were extracted from bacteria and fungi, the most notable example being the discovery of penicillin by Alexander Fleming in 1928.

The Second Epoch: 1945–1970s

This then takes us to the second major epoch in the evolution of the pharmaceutical industry, which began during World War II with the industrial production of synthetic penicillin (Malerba and Orsenigo, 2002). From the 1940s, the industry underwent significant change as large R&D programs became more structured and formalized. The role of scientific drug discovery and experimental medicine also advanced and became central to the commercial business of marketing therapies. The R&D/sales ratio for companies increased significantly during this period, and investment in marketing and sales to doctors, rather than patients, became the norm (Garavaglia et al., 2006). Furthermore, it was during this period that the first major public investments in basic research were made. Coupled with the creation of socialized health care in Europe, this provided both an advancement in new scientific knowledge and research capabilities and a large, single, and well-organized market for new therapies (Garaveglia et al., 2006: 237).

Here, we can observe the beginnings of a diverse innovation ecosystem for therapeutic innovation with multiple value chains, institutions, and actors (both public and commercial) shaping new product development pathways. In the first few decades following World War II, many new products were introduced, which arguably transformed both the clinic and broader society. These included various broad spectrum antibiotics, new vaccines for childhood diseases, tranquilizers and antidepressants, steroids, beta-blockers, and oral contraceptives (Grabowski, 2011: 162). A defining feature of this second epoch was a change in commercial strategy, from one where companies tried to discover markets for products they had developed (the predominant nineteenth-century model), to one where companies designed products for specific markets and then primed those markets for maximum clinical uptake. Nightingale and Mahdi (2006) illustrate this with the example of Sir James Black's work in the 1960s on H2 receptor antagonists, which block the action of histamine on stomach epithelial cells and thereby reduce acid production. Here, deeper understanding of the structure of disease targets was used to improve selection and modification of small molecule drugs. In the 1970s, these insights led to the manufacture of the blockbuster drug Tagamet (cimetidine) for the treatment of stomach ulcers, by the multinational chemical company SmithKline & French. Hopkins et al. (2007) characterize the postwar period as defined by a "synthetic organic chemistry heuristic":

> During the post war period the plant-based extractive and biological traditions waned as a synthetic organic chemistry heuristic provided the pharmaceutical industry with a "golden age" of productivity driven by random screening of synthetic compounds characterised as "molecular roulette." (Hopkins et al., 2007: 568)

Deeper knowledge and understanding of the basic biological path-ways of disease remained limited throughout the 1960s and 1970s, so drug discovery during this period rested largely on serendipity and this "molecular roulette." The promise of rational drug design based on well-understood principles of biochemistry would come later. Le Fanu (2011) suggests that many of the groundbreaking therapeutic discoveries from the 1930s to the 1950s (the steroid cortisone, the antibiotic streptomycin, and the immunosuppressant chlorpromaz-ine, for example) could not have been discovered from first principles. It was very much a trial-and-error process based on random screening and testing in a largely unregulated environment, where a new disci-pline of clinical science was being shaped.

The basic industrial R&D model was to screen known chemical compounds or randomly test in vivo any molecules that the burgeon-ing chemical companies had stored in their vast chemical libraries. This was colloquially referred to as the "shotgun approach." Medicinal chemists would then optimize any promising "lead molecules" that emerged from this process to produce drug candidates, which were then passed down to late-stage development (encompassing the clini-cal trial phases) and eventually to market (Mittra, 2008). A limitation of this approach was that few molecules in these libraries had suffi-ciently high structural diversity to be viable drug candidates. There was also an over-reliance within industry on imperfect animal models that proved to predict human response relatively poorly.

The Third Epoch: Late 1970s–Early 1990s

The third epoch of the pharmaceutical industry gradually emerged in the late 1970s and early 1980s (Malerba and Orsenigo, 2002). During this period, the industry began to transition from an R&D model based on random screening to one based on "guided drug discov-ery" or "rational drug design" (Malerba and Orsenigo, 2002: 669). This required knowledge of the molecular structure of drug targets, which had been missing during the 1960s so-called golden age of drug discovery. Major scientific advances in molecular biochemistry and pharmacology heightened the promise and expectation of more efficient and targeted drug discovery and development in the 1970s and 1980s, as the biological sciences began to take a more pivotal role in drug development. The industry learned much more about drug targets and tried to rationalize the screening process and make R&D more efficient. Many new classes of drugs were marketed in the 1980s based on these advances. Grabowski (2011) notes that many of these therapies focused on specific receptor targets with novel mechanisms of action, and included "interleukin-2 [a type of cytokine signalling molecule needed for immune response] inhibitors to prevent organ transplant rejection, the statins for cholesterol reduction, the nucleoside reverse inhibitors for HIV-AIDS, the serotonin reuptake inhibitor for depression, and proton pump inhibitors for GERD [gastroesophageal reflux disease]" (Grabowski, 2011: 164).

It was also in the early 1980s that the fledgling dedicated biotechnology firms were emerging in the United States, some of which would become large multinational corporations in the 1990s (e.g., Amgen, Genzyme, Millennium, and Genentech). These companies had been at the forefront of the development of recombinant DNA technologies. Genentech, for example, was the first company to synthesize human insulin, in collaboration with a US national medical center. By inserting the gene for human insulin into the deoxyribonucleic acid (DNA) of a bacterium, researchers at the company created the two key protein chains that were then combined using chemical synthesis to produce human recombinant insulin. This was then marketed as Humulin in the 1980s, by the pharmaceutical company Eli Lilly, which had acquired the product. For the first time, diabetes patients did not have to rely on animal insulin, which Eli Lilly claimed was an inferior product. So by the mid-1970s, the technical, industrial, and clinical dynamics of molecular biology had advanced sufficiently to promise imminent application. This took recombinant DNA science away from its esoteric origins within university laboratories, which was helped by the expansion of intellectual property rights following the Bayh-Dole Act in 1980 (Kraft, 2013: 37).[3]

A Changing Innovation Ecosystem in the 1990s

From the end of the war until the early 1990s, the innovation ecosystem for therapeutic R&D was highly concentrated with large, multinational, and vertically integrated chemical firms. Evolution of the industry was defined by "organic growth" as these large companies slowly built up internal research, production, and marketing capabilities so that they could take a product from early-stage discovery all the way to a blockbuster market. These firms financed their own R&D expenditure through the vast profits they made from drug sales and they all tended to hold very little debt (Grabowski, 2011). This was an era in which large chemical conglomerates operated with relative freedom and independence. They were not reliant on the knowledge, expertise, and products of external innovators as everything was conducted in-house (Coombs and Metcalfe, 2002). Having said that it is important to recognize that pharmaceutical firms have historically commercialized many basic science discoveries originating in universities. The noncommercial sector has always been a valuable contributor to therapeutic advance, as I explain in more detail in chapter 4.

The structure of the industry and its innovation ecosystem began to change in the 1990s as R&D was transformed by advances in knowledge about molecular biology, genomics, and synthetic chemistry, as well as the emergence of much faster and more efficient screening technologies based on combinatorial chemistry. Such technologies increased the rate at which new chemical and molecular entities were discovered by companies and furthered their knowledge about disease targets (Ratti and Trist, 2001). Scannell et al. (2011) argue that in the 1980s and early 1990s, combinatorial chemistry increased the number of molecules that could be synthesized by a chemist 800-fold per year. High-capacity screening and better target identification and validation technologies fundamentally changed the nature of internal R&D and the structure and complexity of product pipelines (Drews, 2000). Because of the interdisciplinary and diffuse nature of life science knowledge, no single company or organization could encompass all the capabilities and expertise required to exploit it for therapeutic benefit. This necessitated a network structure for the sector and greater specialization in R&D functions (West and Nightingale, 2009). It also engendered a more diverse set of value chains. Pharmaceutical R&D came to display the characteristics of a "distributed innovation system" (Cambriosio et al., 2004; Chiesa and Toletti, 2004). Here, large firms exploit mergers, acquisitions, strategic alliances, and licensing deals with other innovators to acquire knowledge, technology, and expertise. This is required to sustain innovation and productivity. Firms now had to coordinate an increasingly diverse range of R&D capabilities alongside the "normal" processes of organic growth (Coombs and Metcalfe, 2002). The emergence of these highly complex and distributed innovation networks and systems, where new types of value can be realized, challenges the conventional stereotype of R&D as a strict linear process, which I discuss in chapter 3.

Furthermore, in the late 1990s, the large chemical companies formally separated their pharmaceutical and agricultural divisions. In the 1980s and early 1990s, synergy between agricultural and pharmaceutical divisions made sense, as firms believed economies of scale could be exploited at the discovery level when both sectors were interested in discovering the sources of chemical novelty through functional genomics. However, this strategy was not appropriate or profitable when companies were trying to exploit specific life science applications and markets (Chataway et al., 2004). The commercial opportunities and markets in the agricultural sector were simply becoming too distinct from those in pharmaceuticals (Mittra et al., 2011). It is within this modern era of new biology, and the particular industry structure that solidified in the late 1990s, which has come to define the contemporary innovation ecosystem for pharmaceuticals and engendered both opportunity and challenge for large pharmaceutical firms.

The "Productivity Crisis" and New Challenges

Knowledge and technologies built on new biology took root in the 1990s at a time when the pharmaceutical industry was failing to match the R&D outputs it had sustained for the previous three decades. Despite the promise of new biology—and an idealistic rhetoric of hope that was being mobilized around the notion of life science-driven health care and repositioning companies in relation to their patients, publics, and markets (Bower, 2005)—a number of factors challenged large firms' dominance in therapeutic innovation (Mittra, 2008).

First, and perhaps most crucially, there was the beginning of a protracted decline in R&D productivity despite increasing R&D investment. The 1960s "golden age" of drug discovery was already in decline by the 1970s. This began to engender a sense of pessimism about the future of medicine (Le Fanu, 2011). With the additional problem of product maturity, meaning that all the easy small-molecule targets had been exploited and were no longer protected by lucrative patents, there was a perception of "innovation deficit" (Drews and Ryser, 1996). Since 1996, the number of small molecule chemicals approved by regulators has been in decline, and the number of new active compounds discovered has remained relatively constant ever since. In the late 1990s, companies were simply not generating enough new compounds within their own research laboratories to sustain the growth trajectories demanded by volatile markets (Horrobin, 2001). Scannell et al. (2011) refer to this decline in R&D efficiency as "Eroom's law." This is the reverse of "Moore's law" (Eroom is Moore spelled backwards), which derives from an observation made in 1965 by Gordon Moore, the cofounder of Intel. He described the exponential increase in the number of transistors that can be placed on integrated circuit boards, which doubles computing power every two years. By contrast, Eroom's law suggests that improvements in science, technology, and management have been overwhelmed by powerful external forces, which has limited their impact and significance (Scannell et al., 2011: 191).

Second, and related to my first point, companies were beginning to experience a high failure rate of new compounds, particularly during phase 2 clinical trials, which I describe in more detail in chapter 5. An unacceptable safety and efficacy profile has generally been regarded as the principal cause of phase 2 attrition, which can be as high as 80 percent (Horig and Pullman, 2004; Pardridge, 2003).

Third, the dominance of large firms was also threatened by rising overall costs of drug discovery, due to the need for new, experimental methodological approaches to drug discovery and development; the increasing internationalization of research and its competitive environment, with associated transaction costs; and the increasing demands of regulators and health-care providers (Howells, 2002; Mittra, 2008). In 2003, the cost for a large pharmaceutical firm to bring one product to market was estimated to be 800 million USD (DiMasi et al., 2003), and this figure has been rising annually. The figure is now estimated to be well over 1 billion USD (the calculation incorporates the cost of companies' failed drug programs, so the higher the failure rate the higher the cost per new drug launched).[4] The safety and efficacy requirements from regulators are continually driving up standards and associated costs of doing business in drug development. Furthermore, pressures from health-care payers to limit reimbursement and demonstrate the "true value" of products over existing therapies continue to render pharmaceuticals an expensive and risky business. As one head of Global Sciences at a major pharmaceutical firm stated when interviewed:

> In development, it [productivity and innovation deficit] is absolutely to do with increasing expectations for safety and efficacy data—coming rightfully from the regulatory authorities and society has ultimately demanded much larger clinical trials. Other parts of development like process technology, manufacturing and so on, are actually being done a lot more efficiently in the industry, paradoxically. But the clinical costs have gone sky-high and timing of doing big clinical trials, the logistics, the competition etc. have definitely been a major contributor to increases in development costs. (Head of Global Sciences, Company 2)

There is evidence to suggest that regulation negatively affects R&D performance and the time and cost to get a new therapy to market (Dove, 2003; Hartley and Maynard, 1982). This has hindered the success of new and often smaller innovative biotechnology companies. They simply cannot compete with the large multinational firms, which can withstand the time and cost pressures, and navigate a complex and often sclerotic regulatory system built on a conventional pharmaceutical model (Tait, 2007).

Stakeholder's high expectations regarding the quality and price of new medicines, in the context of controversies over safety and the uncertain therapeutic value of new drug therapies (Abraham and Davis, 2007), also partly contributed to the escalating costs and difficulties of pharmaceutical innovation in the 1990s. Indeed, Abraham and Davis advocate that we distinguish "product innovation" from "therapeutic innovation," because the former does not necessarily entail the latter, and: "...while product innovation retains commercial significance for pharmaceutical manufacturers, irrespective of therapeutic innovation, it is generally therapeutic innovation which is of most value to patients, public health and health professionals" (Abraham and Davis, 2007: 389). However, true therapeutic innovation, as opposed to incremental innovation via "me-too" therapies[5] that large pharmaceutical firms are often accused of prioritizing, is slow, difficult, and expensive. Large companies must carefully balance investments in product innovation and therapeutic innovation to deliver shareholder value, and this tension has been partly responsible for the productivity challenge (Kraft and Rothman, 2008). Linked to this notion has been a long-standing critique and debate around what is termed "pharmaceuticalization," which refers to how industry, in response to the innovation crisis, has sought to identify, or invent, new diseases for which existing products can be targeted (Abraham, 2010; Williams et al., 2013).

Fourth, there was, and continues to be, an unhealthy and unsustainable reliance on the "blockbuster" model of drug discovery, and a reluctance to move too quickly into more targeted therapies for niche markets. The latter would require a much greater prioritization of therapies for unmet medical need. Historically, large firms have targeted three or four key blockbuster therapy markets to remain competitive and sustain revenue growth. Despite emerging hype and hope around the prospect of drug pipeline portfolios built on radically new therapies for smaller populations (such as RM and personalized or stratified medicines, which I discuss in more detail in later chapters), pharmaceutical firms did not aspire to transform themselves into life science companies. Instead, they continued to aggressively pursue the small-molecule blockbuster products with which they were familiar and had the required capabilities and expertise. However, as patents began to expire on many major blockbuster products, and the next generation therapies became increasingly difficult to discover, develop, and market, firms were driven to consider adopting a new modus operandi. They began to integrate new biology more aggressively into systems, processes, and corporate strategies built on traditional chemistry. Styhre and Sundgren (2011) report that many scientists in large pharmaceutical firms came to believe that this blockbuster model based on wet lab in vivo biology research was beginning to transform into a more biocomputational model. Companies were aligning traditional science and new technologies into a new regime of drug development. However, integrating new biology and traditional chemistry proved incredibly challenging, as I describe later. My interviews with senior representatives within large pharmaceutical firms revealed growing skepticism about the blockbuster model of drug development. One respondent stated:

> If you add up the R&D challenges and the R&D productivity issues with our current commercial model you would say the age of the blockbuster is dead because it is not looking to us as though it's sustainable. We've got to come up with a different balance between our commercial practices and our route to market, and also an R&D process that's more accommodating of projects of different sizes and opportunities, perhaps aided and abetted by diagnostics and other things, which allow us to develop them a little bit more effectively. (Senior Portfolio Manager, Company 2)

Nevertheless, other respondents cautioned that although the blockbuster drug's future was uncertain, and the industry was clearly in a precarious transition phase at the turn of the century (due to the productivity crisis), it was not at all clear what would replace these products. Furthermore, how would new business models and value chains be created and made to work for more niche and targeted therapies within the broader innovation ecosystem? All the major pharmaceutical companies continued to embrace a blockbuster drug development strategy, but they were anxious about the future and tentatively explored alternative options offered by the life sciences.

Together, these industry challenges, or series of crises, have shaped the evolution of the pharmaceutical industry and the organization and management of companies' internal R&D. The question I want to address now is how did individual companies evaluate the opportunities and challenges of new biology, and assess new technologies, in terms of their impact on R&D efficiency and the economic bottom line?

New Biology as Both Opportunity and Challenge

Nightingale (2000) has argued that from the early 1990s, biotechnology created new "economies of scale" in early-stage R&D for pharmaceutical firms. High-throughput technologies allowed companies to increase screening capacity with less human resource and thereby reduce the size of the experimental unit. This undoubtedly reduced the overall cost of discovering new drug compounds, while generating valuable efficiency gains. Nightingale suggests that both chemistry and biology shifted from largely craft-based, sequential processes of experimentation on single compounds to automated mass-production processes conducted in parallel. In the 1990s, almost all pharmaceutical companies adopted this common technological platform for discovering new drugs. Advances in combinatorial chemistry meant that random screening could be done with compound libraries sufficiently large and diverse to have a high probability of finding new, therapeutically active molecules. My interviews with senior R&D managers within large pharmaceutical companies revealed the promissory visions and expectations that were generated around new biology in the early days. In the context of the specific challenges firms were facing, new life science technologies emerged as a potential panacea for industry. Most respondents supported the argument elaborated by Nightingale that in the late 1990s life science technologies created economies of scale and efficiency gains. A number of transformative technologies were considered by my respondents to have been particularly favorable to the R&D process.

Of particular value were the automation technologies for combinatorial chemistry, which produced at an exponential rate new compounds that could be tested for pharmacological activity. This represented a major shift from the more traditional craft-based modification of single molecules by medicinal chemists that had prevailed in the preceding decades. One respondent stated:

> Automation has been quite a significant issue…In chemistry I suppose combinatorial chemistry, automated synthesis, has increased the number of compounds made whereas in the past it was more a careful design process of one particular molecule. In biology there have been massive strides in trying to break down the process and the involvement of particular targets, enzymes and receptors in diseases and screening for them in a much more controlled way. Breaking the system down really in kind of a reductionist way and then using automation to try and generate a lot more data. So I think we've had a massive data explosion. (Senior Portfolio Manager, Company 2)

Knowledge and understanding of complex biological structures and disease pathways also benefited from the development of technologies such as X-Ray crystallography and electron microscopy, which respondents felt were particularly important for small-molecule drug development. They allowed organic chemists to design much better molecules that would interact with drug targets in desirable and predictable ways. This was capitalizing on the promise of "rational drug design," which fundamentally changed the organizational structure of the R&D process, as highlighted by Cockburn (2004). It also represented, according to one head of R&D policy at a major pharmaceutical company, a shift from "empirically-driven R&D" to "target-driven R&D." One head of Global Sciences and Information stated that by the turn of the twenty-first century, analytical technologies had advanced beyond all recognition. He explained that his company could synthesize a compound and elicit its molecular structure within an hour, when it would previously have taken over a week. This, he argued, has had a significant impact on productivity. Platform analytical technologies led to major efficiency gains in the screening and selection of viable therapeutic compounds, and it was within this context that they were most highly valued.

The Challenge of New Biology and Changing Industry Expectations

Nevertheless, expectations of new biology and the prospect of a burgeoning health bioeconomy went far beyond the efficiency gains from better screening and selection of new small-molecule compounds. There was also promissory value in developments in DNA sequencing and functional analysis following the mapping of the human genome. This had led to the identification of ever more drug targets for therapy, beyond the 500 or so known drug targets on which industry had built a deep repository of knowledge. Targets now included a new panoply of receptors, enzymes, and large numbers of previously unknown proteins (Dahl and Sylte, 2006). There were now potentially thousands of novel drug targets for therapeutic activity, which promised to generate an endless stream of new therapeutic products and raised industry expectations. However, these expectations could only materialize once the drug targets were validated and molecules could be discovered with the correct pharmacological properties to effectively interact with them. As one respondent put it: "As the technology increased its stronghold in genomics, targets were being turned over very rapidly but were not validated and this significantly impacted their efficiency and their value in a negative manner" (Former Head of Computational Biology, Company 3). A former senior scientist in bioinformatics also pointed out:

> To validate the target you need more information, it's not like pressing a button. You need to have other types of data, and you need

to integrate other types of data with sequence data. There is a major problem in trying to get comparable results...you need to design and develop new methods and algorithms for data integration and analysis, and so called hypothesis generation, so you can rank targets using software and make plausible guesses as to which type is better. (Former Senior Scientist in Bioinformatics, Company 1)

Without the ability to validate rank drug targets in a meaningful way, new biology would not be a magic bullet for the pharmaceutical industry to use on the challenges it faced at the turn of the century. Cook et al. (2014) suggest that the industrialization of discovery research led to productivity being defined by quantity-based metrics, and the assumption that increases in compounds discovered would correlate with an increase in the number of new products launched. However, this did not happen and the authors argue that the very culture of research and organization in companies changed in a negative way. Volume-based goals replaced the more valuable curiosity-driven drug research.

At the time that genomics and biotechnology were heralded by many scientists and the popular press as a revolutionary new stage of therapeutic innovation, and potentially transformative of health care, some academics in innovation studies questioned the evidential basis and intellectual edifice supporting such claims. In a controversial but highly influential paper, Nightingale and Martin (2004) challenged the arguments of academics, industry, consultants, and government that society was in the midst of a biotechnology revolution. The paper was a powerful critique of the promissory rhetoric that investments in life science were improving therapy development and contributing to the health bioeconomy. This argument was later reiterated by Hopkins et al. (2007) in a more detailed analysis. The basis of their argument was that medicinal biotechnology had failed to engender revolutionary change in R&D outputs. Instead, it was following a conventional pattern of slow and incremental technology diffusion. According to these authors, expectations built around new biology at the turn of the twenty-first century were widely overoptimistic. The corollary of this argument was that policymaking assumptions about the role of biotechnology in the health bioeconomy needed to be revised.

Many of my interview respondents shared this skepticism about the downstream impact of new biology on drug development, so we should not assume that the hype was being primarily generated from within the industry. Although most respondents were sanguine about the impact of new platform technologies on early stage drug discovery, they were cautious about making bolder claims for revolutionary downstream impacts in the clinic. This was largely due to the continuing challenges new biology posed to an industry that had built its knowledge, capabilities, and markets around simple, small-molecule drugs. One respondent stated that identification and selection of promising drug candidates had been improved by new technology, but: "If you ask me have we got anything to show for it in terms of new drugs, I'd be fibbing if I said we have" (Director of Academic Liaison, Company 1). Another respondent was more explicit in stating that expectations of imminent benefit from new biology and data from the Human Genome Project were premature. He argued that because of the fact that the human genome turned out to contain fewer genes than expected, this meant industry had to revise its estimates of the size of the druggable genome; that is the number of proteins belonging to particular structural classes for which small-molecule ligands are known. He estimated this to be around 10 percent and stated pessimistically that the druggable genome has probably already been mined for small-molecule drugs (Former Head of Neuroscience, Company 4).

Bunnage (2011) suggests that the challenge is not only a narrowing of drug target opportunities, but equally important is the difficulty in selecting those targets that will effectively modulate disease. Nevertheless, the comments of the respondents above support the point I made earlier about the maturity of product pipelines and the increasing difficulty of discovering and marketing truly innovative therapies—hence the saturation of the market with incremental "me-too" therapies. Part of the problem is that new biology is different and perhaps unique in ways that other sciences are not, as I explained in chapter 1. Unlike most nonbiological disciplines, such as engineering, biological systems are complex, diverse, and dynamic in a way that makes it difficult to formulate generally applicable laws. Uncertainty and serendipity is a defining feature of biology and knowledge is advanced by radical breakthroughs rather than incrementally (West and Nightingale, 2009: 559). Biology projects cannot be modularized and managed in the same way as other types of science projects. So the shift from small-molecule drug development, based on well-defined and concentrated knowledge and expertise built up over decades, to a new paradigm of drug development rooted in new biology, is a long and punctuated process of success and failure. Furthermore, as I show in subsequent chapters, the external constraints on life science innovation, such as capricious regulatory systems, changes in market expectations, and uncertain health-care pathways and reimbursement policies, shape options for product development and determine their commercial, economic, and social value.

A number of my interview respondents also talked about the challenges of organizing people and processes around new innovation paradigms, and how expectations often conflict with the reality of what is possible. One respondent stated:

> In the 1990s George Poste, who was at that time head of research of Smith Kline Beecham before the merger with Glaxo Wellcome, was a great advocate of genomics; he went around saying it's going to transform everything. They put their money where their mouth was in a very big deal with Millennium [a biotechnology company founded in 1993]. There was a lot of feeling around that time that genomics was just going to be absolutely marvellous. I think most people in the industry would agree that it has not fulfilled certainly the most extreme promises, and perhaps even a measured view would say it's been pretty disappointing. You can adduce a number of reasons. The general climate of stricter regulation. But also the interdisciplinary point of how easy it is for people in large research bureaucracies to operate in an environment which is different from what they're used to. The old way of the pharmaceutical industry was that medicinal chemists ruled...the driving thing was the imagination of medicinal chemistry. The molecule benders were the important people really. Now that is changing. (Former CEO of a Biotechnology Company)

This challenge of organizing R&D to integrate new, interdisciplinary knowledge and processes with existing and well-entrenched capabilities is important in understanding the evolving relationships between pharmaceutical companies and the smaller biotechnology and genomics companies at the turn of the century. How did large pharmaceutical firms begin to acquire knowledge, technology, and expertise in new biology and attempt to integrate it with drug pipelines built on small-molecule drugs?

One of the defining features of the pharmaceutical industry at the dawn of the twenty-first century was the increasing value it placed on merger, acquisition, and strategic alliances. This had a material effect on the broader innovation ecosystem within which the industry operated. The strategic behavior of the largest firms during this period foregrounds many of the transformations in organizational and institutional practices that were later driven by the expectations and demands of the new health bioeconomy. Within this vibrant but fragile innovation ecosystem, the relationship and tension between large, incumbent pharmaceutical firms and new, innovative life science companies was a key dynamic that drove knowledge and technology flows.

The Dynamics of Knowledge and Technology Acquisition

So far, I have revealed how new biology challenged conventional pharmaceutical innovation processes, but also presented new opportunities as industry responded to crises around productivity and perceived innovation deficit. Companies promiscuously adopted different strategies to extract value from nascent molecular screening technologies and new biological science to complement existing capabilities in small-molecule drug development. In this section, I explore the distributed innovation ecosystem that emerged partly as a result of this new biology and the expectations of value that built up around it. In particular, I look at this through the lens of merger, acquisition, and strategic alliance behavior. I explain the factors that drove merger and acquisition (M&A), strategic alliance, and licensing behavior, and reveal how large companies sought to balance in-house R&D with securing knowledge and technology being developed in the smaller biotechnology sector. These industry trends provide insight into large companies' ability and commitment to acquire and exploit capabilities in new biology. They also reveal how industry practitioners ascribed different types of value to new technologies and rationalized their corporate strategies.

Industry Consolidation through Large-Scale M&A

Historically, large company mergers have tended to cluster by industry, and certain periods have been defined by particular types of merger (Allen et al., 2002). So "conglomerate mergers," where acquiring firms build up a diverse group of companies to form major conglomerates, defined much of the 1960s. "Bust-up" takeovers, in which the conglomerates that were established in the 1960s were broken up and parts sold off by acquiring companies, defined the 1980s. In the 1990s, there was a return to significant industry consolidation, which was similar to the pattern of M&A that defined the 1920s. In the 1990s, all the major pharmaceutical firms countered each other's strategic moves by building capabilities in life science and pursuing major M&A activity. This phenomenon is nicely explained through the concept of "institutional isomorphism," which DiMaggio and Powell (1983) coined to describe the process through which large organizations, acting as rational agents, become indistinguishable from each other as they tend to deploy similar strategies in response to risk and uncertainty. Maturing markets for conventional small-molecule drugs, and uncertain pathways to the clinic for new biological therapies, precipitated this prolonged period of large-scale M&A. In the period 1990–2004, there were 22 major pharmaceutical M&As, with 5 exceeding 60 billion USD (Sandoz and Ciba-Geigy merger in 1996; Pfizer's purchase of Warner-Lambert in 2000; Glaxo Wellcome and SmithKlineBeecham merger in 2000; Pfizer's purchase of Pharmacia in 2003; and Sanofi's purchase of Aventis in 2004). Only two of the top ten pharmaceutical firms that were operating in 2004 (Merck & Co and Eli Lilly) had retained their position through organic growth, whereas the remaining eight had been involved in M&A activity to varying degrees, particularly the European-based companies (Mittra, 2007). All but two major European pharmaceutical companies merged or were acquired between 1997 and 2002, when the promissory claim that new biology would revolutionize the industry was at its apotheosis.

Many factors drove industry consolidation, and they partly reflect the challenges I described earlier. They included the expiry of patents on lucrative blockbuster drugs; the need to build critical mass and curtail inefficient processes and duplicated activities (companies sought value from new economies of scale); and the desire to expand into new and emerging global markets, so companies would buy competitors if, for example, they had personnel and resources in key target markets (Mittra, 2006, 2007). It is noticeable that many of the drivers for M&A in the pharmaceutical industry have been negative in character. They have been used by companies in response to perceived weakness or risk in their product pipelines. So companies have looked at their drug pipeline and estimated its current and future value, and then sought to pursue M&A to de-risk their portfolio and build a sustainable future. One interview respondent described large mergers as driven by the need for sustainable profit in the context of decreasing prices and increased costs of development. Companies might look at another company with a complementary product and reason that through a merger both products could be taken to market at a much lower cost (National Projects Leader and Director of Business Development, Company 2). In this case, the merger can allow the company to better utilize its capabilities and resources. However, there is a potential downside, as diseconomies of scale can result from the costs associated with managing increasingly large and geographically distributed research groups (Henderson and Cockburn, 1997: 10–11). Scaling up, in this case, undermines a company's ability to make best use of its internal capacity. Furthermore, innovation and productivity can actually reduce as the merged company tries to reduce its costs by curtailing its research programs, such that innovative energy and support for experimentation is replaced with a need to standardize operations and maintain the status quo (Berggren, 2001: 14). Here, consolidation can stagnate innovation, at least in the short term.

The literature and the evidence from my interview respondents suggest that M&A can engender certain benefits in terms of opening up new markets and access to new products and technologies, but long-term sustainable growth can be threatened by the pressure to discover ever more blockbuster therapies (Graves and Langowitz, 1993). That is, as these companies grow ever larger, they must increase their productivity and output accordingly to sustain profit levels. Indeed, there is evidence to suggest that, in the long term, merged companies do less well, in terms of operating profit three years post-merger, than companies that do not merge (Danzon et al., 2004). Furthermore, the fact that much of the innovative work in new biology is conducted within small and medium-sized biotechnology companies, which are now important and increasingly valued players in the broader innovation ecosystem, suggests that size is not the only requirement for innovation and productivity, particularly for basic science where diseconomies can result in organizations that are too large, unyielding, and bureaucratic. Having said that, there are other benefits to being a large corporation, and this includes the ability to take advantage of "economies of scope" (Henderson, 2000). What this means is that large companies, because they have such a diverse range of research projects, benefit from knowledge spillovers and the ability to exploit assets, such as technology and human resource, in multiple applications without incurring additional costs (Mittra, 2007). So, from this perspective, large firms generate higher value from research activities than smaller firms because they are more likely to generate and be able to exploit these important knowledge spillovers. Also, because these firms house such a diverse range of technologies, skills, and expertise, they have a high level of what Cohen and Levinthal (1990) describe as "absorptive capacity." This is the ability to understand and value new technologies and knowledge, such that they can be absorbed successfully into the organization. One interview respondent captured the essence of "absorptive capacity" when he stated: "If you don't understand it [a new technology or therapy] and are not enmeshed with the people working in the area, you're certainly not going to know what to do with it" (Former Research Scientist, Company 3).

However, despite recognition that there are some scale advantages to being a large, multinational pharmaceutical firm, there was a feeling during the time I was conducting interviews with scientists and managers working in large companies that further M&A was unsustainable. Although there was an obvious financial advantage of merging two companies, in that large expenses could be written off and inefficient operations closed down (Former Head of Computational Biology, Company 3), some felt that most of the sensible mergers had happened and any more would be driven largely from desperation. One respondent stated: "These companies have got to a size where if you make them any bigger it won't work. I think it's marginal whether they're working now, because they're so big" (Research Scientist, Company 1). Nevertheless, despite this growing skepticism, further industry consolidation through M&A continued apace, and continues to do so today. We have seen this more recently with Pfizer's purchase of Wyeth in 2009 for 68 billion USD, and its attempt, which ultimately failed, to purchase the British-Swedish company AstraZeneca in early 2014, although the latter was likely driven by specific financial and tax advantages.

It is important to note that in the first few years of the twenty-first century, the very few large biotechnology companies that had been growing since the early 1980s began to adopt M&A strategies similar to the traditional pharmaceutical firms. For example, Amgen purchased Immunex in 2003 for 16 billion USD (to acquire the potentially lucrative arthritis drug Enbrel), and Biogen purchased Idec in 2003 for 6.8 billlion USD (to acquire a range of cancer drugs). These firms were attempting to supplement their in-house R&D pipelines and diversify their technological and product portfolios. They were adapting their organizational structure, internal systems and processes, and capabilities in light of an innovation ecosystem still dominated by the development of small-molecule compounds (Mittra, 2007). An example of this was Amgen's 2003 purchase, for 1.3 billion USD, of Tularik, which was a small pharmaceutical discovery and gene regulation company, to acquire its discovery capabilities in small molecules and pipeline of diabetes products.

However, the fact that so few dedicated biotechnology firms have successfully grown into major multinational companies, and none since the 1990s, suggests that therapeutic innovation is still dominated by the old pharmaceutical companies. One reason for this is that the regulatory system is built on a conventional, small-molecule drug model, which makes it almost impossible for smaller firms specializing in therapies based on new biology to navigate without the help of a large multinational (Mittra et al., 2015; Tait, 2007). Hopkins et al. (2013) also point to changes in the funding environment and the expectations of investors, which have constrained companies by compressing their development life cycles and reduced their ability to generate late-stage drug candidates that are valuable to large pharmaceutical firms. So while the smaller biotechnology sector is important in the broader innovation ecosystem, its progress and ultimate value is largely determined by the strategies of the traditional pharmaceutical firms.

Nevertheless, restructuring of the pharmaceutical and biotechnology industries continued as a result of various waves of M&A activity, and 1990–2004 was a significant period for this kind of activity. The diversity of firms with highly specialized capabilities within the innovation ecosystem was also reshaping the operating environment for pharmaceutical innovation. Large pharmaceutical companies were confronted by a growing biotechnology sector that was building the technologies, knowledge, and expertise to develop twenty-first century therapies based on the life sciences. The large pharmaceutical companies now had to pursue a variety of strategic options, in parallel, to try to capture what they considered the most valuable biologics-based knowledge, expertise, and products to maintain a competitive advantage and create sustainable growth.

One significant and highly lucrative strategy was to acquire small biotechnology companies with technology or products that the large firms did not have in-house. So these had a very different rationale from large-scale mergers in that they were, as one respondent claimed, driven predominantly by the technology and products, rather than financial factors (Head of R&D Policy, Company 1). Another respondent argued that the key driver for acquiring small companies was: "...to get another piece of technology which is going to help our main cause, and actually integrate it within our organisation so it helps drive our science" (Director of Academic liaison, Company 4). So, multinational companies tended to view the growing biotechnology sector as a valuable source of new innovation that could supplement internal R&D efforts and allow new capabilities to be built (Mittra, 2007). In what was an uncertain and evolving innovation ecosystem, where conventional small-molecule, blockbuster therapies were co-evolving with path-breaking biological therapies, large firms pursued small companies to quickly and efficiently acquire new knowledge, technology, and potentially valuable products. By purchasing these companies, they were creating opportunities along new value chains and adopting new valuation metrics and practices more appropriate to these new types of products.

Monoclonal antibodies provide a nice illustrative example of how large firms adapted their strategies and notional ideas of the value of new path-breaking therapies. These antibodies became of interest to the pharmaceutical industry in the mid-1990s and drove many small-scale acquisitions. Monoclonal antibodies, which were pioneered by Georges Kohler and Cesar Milstein in 1975, are monospecific (have an affinity for the same antigen) antibodies cloned from a single ancestral cell (so they are all identical) that bind to specific antigens and confer a therapeutic benefit. They were among the first life science-based therapies widely adopted by large pharmaceutical firms and fully integrated with their small-molecule pipelines.

In the 1990s, many large firms sought to acquire companies with monoclonal antibody technologies and products, or which had valuable tacit knowledge and understanding of this highly specialized field that was predicted to be a major growth area. This optimistic speculation about future value turned out to be true, as a large proportion of new drugs today are monoclonal antibodies and they are showing significant success in the clinic (Nelson et al., 2010). One notable example of such an acquisition was AstraZeneca's 2006 purchase of Cambridge Antibody Technologies, for 1 billion USD. Cambridge Antibody Technologies developed the technology which led to the first fully humanized blockbuster drug Humira (adalimumab)—a novel anti-inflammatory therapy for arthritis. The deal provided AstraZeneca with access to a whole range of antibody technologies and new drugs. This example, which is just one of many, represented the continuation of a broader trend, which saw large pharmaceutical firms build capabilities in technological niches through acquisition. As we shall see in the following section, acquisitions often followed both formal and informal collaborations and alliances (Mittra, 2007).

Balancing Internal R&D and External Alliances

The M&A strategies that led to industry consolidation in the late 1990s and start of the twenty-first century, and which saw many innovative small companies bought up, were not on their own a curative for the pharmaceutical industry's productivity crisis. Therefore, strategic alliances, collaborations, and licensing deals (product or technology is licensed for a fee, rather than purchased) were also aggressively pursued by pharmaceutical firms. This had a material impact on the innovation ecosystem. Some authors claim that the interdependent networks that were emerging from multiple alliances and collaborations represented a new organizational form (Lukkonen, 2005). Over 1,500 alliances were formed by the top 20 pharmaceutical companies between 1997 and 2002 (Lam, 2004), which highlights the value firms began to place on these kinds of activities. However, it is important to recognize that strategic alliances can be very diverse in character, and there were a variety of ways in which individual companies began to exploit them in conjunction with in-house R&D. Different strategies engendered different levels of value for the pharmaceutical firms that adopted them.

The Value of Alliances as an Exploratory Tool

Alliances with smaller biotechnology and genomics companies were valued by pharmaceutical companies in terms of the options they opened for exploring or "dabbling" in new technologies. As the everyday practices of modern drug development changed in response to the emergence of the disruptive life sciences, there was an increasing dependency of the multinationals on innovation emerging from this vibrant sector. The following account highlights the key reasons for forming alliances and the speculative value judgments that underpinned this strategy:

> We're changing not only the diseases that we're going for but we're changing the way in which we are working, and we cannot afford to bring everything in-house and try it ourselves. The most cost-effective way for us to explore new ways of working or new technologies is to collaborate with a smaller company or academic group and then, when it's successful, bring it in-house. In some cases the expertise is so specialized that we probably wouldn't want it in-house anyway, because we might not need to use it forever... The sorts of collaborations we go in for nowadays are either in areas where we have no expertise at all, like monoclonal antibodies, or it's a very precise area where we see future potential. (Vice President of R&D Science Policy – Company 2)

In this context, there is a degree of institutional or organizational learning at the heart of strategic alliances. Large companies can learn through their collaborations, in a relatively informal way, and build capabilities slowly. These capabilities, and often tangible technological assets, may later be brought in-house if the company feels it can extract greater value through this strategy. This is far less costly than M&A and also provides a degree of flexibility that is highly valued in the context of an ever-changing and sometimes turbulent innovation ecosystem. This aspect was nicely captured in the following account:

> Our two major collaborations in the last 18 months [speaking in 2005] with monoclonal antibody companies fall under the category of

big-small technology deal. We haven't got internal monoclonal inven-
tion capability for humanised monoclonal antibodies so we deliber-
ately went out to seek a major collaboration that's flexible and jumps
on the technological advance of those companies. It obviously costs
you a little bit more than if you invented it yourself but the telescoping
of time and the creation of a realistic portfolio within a couple of years
is fantastic compared to what we would have done if we'd grinded
through. (Head of Global Sciences and Information Company 2)

The fact that large companies pursued these kinds of external
strategies for extracting capabilities and know-how in new biology
suggests they recognized that the conventional pharmaceutical R&D
model was broken in some fundamental sense. It also implied, as
Drews and Ryser (1996) suggest, that large firms lacked the flexibility, originality, and innovative lateral thinking required to develop
and derive value from new biology. This chimes with Nightingale and
Martin's (2004) argument, which I discussed earlier, that large pharmaceutical companies lack the capacity for change, as evidenced by
their inability to fully develop biotechnology in-house and deliver on
the promissory visions and expectations of truly innovative therapies
based on this biological paradigm. Instead, they have had to look for
new innovative capabilities within the evermore complex networks of
alliances within a much broader and more expansive innovation ecosystem. Nevertheless, it is important to recognize that large firms do
have the capacity for change. The fact that they did not fully embrace
biotechnology in the early days, and integrate the new therapeutic
paradigms alongside conventional small-molecule drug development,
perhaps reflected their strategic priorities at the time. Firms must continually make strategic choices in response to changes in their operating environment, and they do this by evaluating time constraints,
cost-effectiveness, and long-term value realization.

If we turn again to the example of monoclonal antibodies, many
of the formal acquisitions of small monoclonal antibody companies
were preceded by different types of formal and informal alliances. At
the same time, many of the smaller partners were themselves involved
in alliances with similar-sized companies, which created a complex
network of innovators contributing to the overall value chain (Mittra,
2007). The decision to formally acquire an alliance partner suggests
that the company had reached a critical stage of maturity in the field.
By bringing the tacit and codified knowledge, expertise, and technological assets from these alliances in-house, the companies were
signaling their confidence in the value and growth potential of the
field (Mittra, 2007), as the quotation above illustrates.

However, in time companies began to recognize that not all alliances are created equal in terms of what they can deliver with regard to short-term and long-term value. Companies moved from being overly optimistic in their speculative assessments of the future value of new technologies, to being far more cautious, as captured in the following account.

> We want to have a very clear business case between what the promise of the innovation is and how it will solve some of our core problems...We made some big brave throws by investing in [various genomics companies] in the early stages...we were more driven by vision than we were by the factual realization of those activities. Now you find us more cautious. (Head of Global Sciences and Information, Company 2)

Here, the value of an alliance is inextricably linked to a set of tangible problems in the large company's R&D pipeline. The value (in terms of generating profit, creating efficiency, or contributing new knowledge) of an alliance with a small, innovative company is not fixed or transcendental, but emerges from day-to-day practices and is often ephemeral. The types of research alliances and collaborations exploited are also often stratified by therapeutic area, so alliances will often cluster around particular fields of study where innovative new approaches from life sciences are thought to be most relevant. Cancer and infectious diseases, for instance, are two areas that historically generated a lot of alliances (Mittra, 2007).

The Shifting Value of Licensing

In the early twenty-first century, pharmaceutical companies increasingly began to acquire new therapeutic compounds through licensing. This represented a further strategy to capture externally sourced innovation and escape the productivity trap. Licensing was considered to be a highly valuable strategy as it enabled companies to cherry-pick what they considered to be desirable compounds, without having to undergo the time and expense purchasing the entire firm. Indeed, there is evidence that therapeutic compounds licenced-in to companies' R&D pipelines tend to have a higher rate of success (Pammolli and Riccaboni, 2000). Although time and money must be spent evaluating the product, the potential pay-off once the product is "internalized" is usually significant.

From the mid-1990s, all the top 20 pharmaceutical firms exploited licensing strategies to sustain their profitability and meet the high expectations of shareholders and the markets. However, there was significant variation in the type, range, and scope of licensing deals, with some companies far more reliant on licensing than others. The European companies tended to give a higher priority to licensing than their US counterparts, with the latter continuing to focus on their internal pipeline (Featherstone and Renfrey, 2004). As one interview respondent stated:

> I think principally we've been developing our own portfolio largely because we have had quite a lot at early-stage that has looked very promising, but lately we've had some difficulties with some late-stage stuff, so I think the strategy's changed a bit and we are going to be looking more actively at in-licensing. (National Projects Leader, Company 5)

However, it is important to recognize that there is a significant risk in licensing-in products to fill a gap in the company's product portfolio, particularly if the company does not have the knowledge and expertise in-house to adequately evaluate the product's value and future potential. The scale and scope of licensing strategies and their ultimate success depend on existing R&D capacity and financial status. As another interviewee nicely put it: "Licensing-in depends on where you are and how desperate you are, and there is no ideal balance between in-house and licensed-in products" (Vice President R&D Science Policy, Company 2).

The balance between formal M&A, alliances, licensing, and internally driven R&D has always been a challenge for strategic management. The elusive quest for rationally-driven decision making is problematized by the risks and uncertainties that are a defining feature of innovative drug R&D, as well as the vagaries of the broader innovation ecosystem. It is within this constant state of flux that the present and future value of licensed products is continually changing and driving companies in a number of strategic directions. The result of this at the turn of the twenty-first century was that the pharmaceutical industry began to look less like the homogeneous behemoth it had been for many decades, with some firms beginning to significantly differentiate themselves in terms of strategy and organization. One significant development that impacted on the pharmaceutical value chain was the increasing competition around the licensing or external purchase of early-stage and late-stage drug compounds, both of which have always been highly valued. Products in the preclinical and clinical phase 1 stages of drug development are in high demand because they are relatively cheap, as their chance of success is largely unpredictable (many will indeed fail). Late-stage products (those that are in phase 3 clinical trials) are also valuable because they have a high probability of being approved by regulators, but they are also much more expensive than early-stage products (Mittra, 2007). Historically, pharmaceutical companies considered drug candidates in mid-stage development to be the least commercially attractive, because they fall within that part of the innovation process where there is a high attrition rate, and it is very difficult to value such products and place an appropriate price on them. However, companies aggressively pursued these more risky products from the late 1990s, because the competition to license the more lucrative late-stage and preclinical products became far more intense and drove prices up. Therefore, many deals were now being made on products in phase 2 clinical trials (Wood Mackenzie, 2004). This is a good example of how changes in the market compelled large companies to reassess their licensing strategies, general risk calculus, and assessment of value.

Finally, in valuing a licensing deal, one might reasonably expect companies to prioritize path-breaking products, which potentially unlock new markets or reconfigure conventional value chains, over those products that represent a minor, incremental improvement or slight modification over an existing therapy. Therapies for cancer, for example, are generally valued more highly if they have a significant impact on mortality rather than an incremental impact. Similarly, biological therapies that promise to redefine health-care pathways, and perhaps revolutionize medical practice, might be considered more valuable than conventional "me-too" therapies (Mittra, 2007). However, the paradox, which is described by Arnold et al. (2002), is that many leading industrialists, investors, and companies do not consider the innovative potential of a therapeutic product to be particularly critical in determining a licensing deal's value. In fact, because there is a great deal of risk and uncertainty involved in identifying and exploiting new markets for highly innovative and path-breaking products, these kinds of products may be deemed less valuable (Arnold et al., 2002: 1087). Interestingly, the authors discovered that if the potential market was large, such as cardiovascular disease, the type of therapy (small molecule or biologic) did not seem to be a particularly significant factor in the licensing decision. This suggests that pharmaceutical companies would be more willing to take on a risky biological product, which may have a long and difficult product development pathway, if the potential market was large.

Conclusion

In this chapter, I have explored the history of pharmaceutical innovation, the challenge and opportunity of new biology, and the changing industrial landscape for pharmaceutical R&D as large multinational companies were compelled to exploit new strategic options for appropriating knowledge, expertise, and products. M&A and strategic alliances provided new options for a highly profitable industry that had been confronted with major technological shocks, a perception of innovation deficit and productivity crisis, and challenges to its historic commercial dominance.

The impact of new biology on pharmaceutical companies and general industry dynamics has been significant. At a time of perceived innovation deficit, biotechnology and genomics emerged as a potential panacea for industry. However, new biology was a double-edged sword for traditional pharmaceutical innovation. On the one hand, it could complement existing drug discovery and development capabilities. On the other hand, it engendered new competitor companies with knowledge and products that required different business models and pathways to market than conventional drugs. The restructuring of the pharmaceutical industry was driven in large part by the uncertainty, complexity, and promissory visions of these life science-based technologies and therapies. Large companies exploited the distributed innovation system to acquire the dynamic range of capabilities now required to profit from the biomedical paradigm, and try to secure new sources of future value on which to sustain the business. Today, they are continuing to face the same challenges and are responding with a similar strategy of consolidation through M&A. They are also continuing to pursue the blockbuster small-molecule drugs.

In the following two chapters, I explore broader changes in the innovation ecosystem, and the emergence of "translational medicine" as both a commercial and broader public policy strategy to deal with various challenges facing therapeutic innovation. Together, these challenges represent what I call the "broken middle" of health innovation. I discuss how R&D has been refashioned by new public policies and commercial strategies, creating what looks like a new institutional landscape for health innovation, comprising multiple value chains, valuation practices, and orders of worth.

Chapter 3

The "Broken Middle" of Health Innovation

Introduction

So far, I have explored the history of pharmaceutical innovation and the challenges and opportunities presented by new biology from the perspective of large pharmaceutical firms. It was important to focus initially on these multinational companies as they have historically dominated therapeutic innovation and defined the patterns of organizational relations and value chains within the health innovation ecosystem. However, in the twenty-first century, the challenges of health innovation and the needs of the growing bioeconomy have led to a much greater role for other actors, institutions, and organizations in therapeutic research and development (R&D). In this chapter,[1] I focus on what I label the "broken middle" of the health innovation pathway. This has led to the emergence of "Translational Medicine" (TM) as a broad organizational strategy to deliver the health and wealth promises of new biology. For the first time, industry, government, and the scientific and clinical communities coalesced around a number of challenges in the middle stages of drug development, which they believed to be responsible for the high failure rate of new therapies. They then sought to establish new, collaborative solutions to the problem. A key concern was the widening gulf—cultural, professional, disciplinary, and epistemic—between basic science in the laboratory and patient care in the clinic. This became expressed through the populist metaphor "the valley of death" (Nature, 2008) and has culminated in a reorientation of research policies, funding priorities, R&D practices, and organizational strategies across the whole health innovation pathway. In both this chapter and chapter 4, I trace these important changes and their material impacts.

Over the past 15 years, the concept of TM has become ubiquitous within health-care and research policy discourses. It is often treated as indispensable if economic and clinical value from new biology is to be fully realized. TM is employed as a broad and often ambiguous trope to describe new ways of organizing and funding R&D activities in the biosciences, and it presages a future of improved patient care. It is also used as a term to describe more tangible new approaches to improve drug development and clinical practice, such as the use of biomarkers and associated diagnostic testing.

The stimulus for TM's arrival, and its increased visibility in various science, policy, and clinical documents and reports (FDA, 2006; Cooksey, 2006; MRC, 2008), was the problem that insufficient novel therapies successfully move from the laboratory to the clinic. Despite unprecedented investment in life sciences following the Human Genome Project (HGP), and the growth of scientific knowledge around the cellular and physiological mechanisms of disease, many promising therapeutic products continue to fail in phase 2 clinical studies. Furthermore, few new therapies that do make it through an incredibly onerous regulatory system become widely adopted as the clinical standard (Milne, 2009). Fewer still successfully breakthrough or transform existing health-care pathways. What I call the "broken middle" of therapeutic R&D has emerged as a powerful "problem narrative" that the pharmaceutical industry believes is rooted in the complexity, risk, and uncertainty of these phase 2 clinical studies. The response has been unprecedented investment from both commercial and public sectors in various translational activities. These have included individual projects, funding mechanisms, and new R&D units/collaborative partnerships to ostensibly improve health innovation.

In chapter 4, I address the practical impacts of TM on interdisciplinary practices, organizational and institutional norms, and R&D processes through some exemplar case studies. The key question I want to address in this chapter is: *What perceived challenges, opportunities, and practitioner values in health innovation have driven a new translational policy agenda, and with what consequences for the bioeconomy?* To answer this, I first explore in some detail the origin of the "broken middle" narrative, and how it has come to drive a complex and multifaceted TM agenda. I then explore the National Institutes of Health (NIH) in the United States as a case example of an anticipatory organization for TM, and reflect on similar initiatives within Europe. This illustrates the important role public sector institutions and policies play in shaping contemporary health innovation options and the values that define them.

I then describe different practitioner perspectives on the nature, role, and value of TM in the context of the perceived challenges to the health innovation ecosystem. These challenges have led to a greater alignment of public sector strategies and values with that of industry. In this context, the boundaries between the commercial and public spheres are becoming increasingly opaque. So how have diverse practitioner and stakeholder views been mobilized to frame the problem of a broken health innovation system, and what assumptions do these narratives make about the relationship between basic and applied science, and the value of TM as a potential solution? Using the current interest in molecular biomarkers as an illustrative example, I reveal both the benefits and limitations of translational approaches and the different types of value that key stakeholders enact and mobilize when discussing the future of new therapies.[2]

The Basic/Applied Research Distinction

The belief in a broken middle of health innovation presupposes a particular view of the historical relationship between "bench and bedside"[3] and the very structure of the health innovation pathway. The assumption of linearity has shaped certain expectations and hopes around TM and come to define its potential scale and scope. Kraft (2013) usefully points out that the ostensibly tenuous relationship between the laboratory and the clinic has long been viewed by policymakers as a barrier to therapeutic innovation, and has become a key target for interventionist strategies from all sectors involved in health innovation. The philosophy of TM, and the promissory visions it has enacted, has emerged as a central policy strategy to drive improvement of the broken innovation cycle (Mittra and Milne, 2013). However, many have questioned the novelty and underlying assumptions of TM, and the purported notions of value and benefit that resonate within biomedical research and clinical communities (Birch, 2012; Birch and Tyfield, 2012; Martin et al., 2008). Furthermore, the very idea of translation, coupled with the rhetoric of a broken R&D system, implicitly assumes a fixed and linear health innovation pathway and clear demarcation of basic and applied research. The hopes and expectations that have been built around new biology and TM may indeed be based on unrealistic or untested assumptions about science, technology, and the nature of R&D, as has been robustly argued in the case of gene mapping (Terwilliger and Goring, 2009).

Stokes (1997) provides a powerful critique of the traditional linear model of innovation, and the conventional distinction between basic and applied research. Historically, basic research was defined as intellectually driven with no specific application in mind, whereas applied research was conceived as a purely practical endeavor. This view came to define science and technology policy in the postwar period.[4] Stokes challenges the underlying assumption of this philosophy with a number of historical examples, including Pasteur's work in microbiology, which he describes as being simultaneously basic and applied research. He argues that Pasteur was not solely committed to understanding the underlying nature of the microbiological processes that he discovered. He also had a strong commitment to control the effects these processes had on animal and human health. Turning to more modern developments, Stokes writes:

> Certainly, the modern biological sciences are difficult to bring within the traditional, either-or view of basic and applied research. The revolution in molecular biology has posed questions, such as how interferon works, that were enormously important both for the advance of fundamental knowledge on recombinant DNA and for major applications— some of which will be immensely profitable. (Stokes, 1997: 14)

There is evidence in a number of related fields that the prevalent view of therapeutic R&D as linear and one directional rarely reflects the reality on the ground. For instance, Martin et al. (2008) note that, historically, the application of basic science was never the caricatured one-directional process often presented in contemporary accounts. The authors cite Lowy's study of cancer therapies using interleukin 11,[5] which required significant contributions from both clinicians and patients during development of the therapy (Lowy, 1997). Lowy's book also highlights the broader sociocultural aspects of biomedical innovation, in this case the clinical trial of a new therapy on a French cancer ward. The distinction between the laboratory and the clinic was far from obvious in this case—the culture of clinical experimentation shaped the development of the therapy—and there were no clearly delineated disciplinary or professional boundaries.

Another much earlier example of the blurred boundaries between basic and clinical research is Banting and Best's 1922 discovery of insulin as a treatment for diabetes. This research moved back and forth from animal models to first-in-human studies and involved many of the interdisciplinary and cross-sector collaborations that are often now heralded as the cornerstone of TM, particularly the close links and sharing of knowledge and expertise between basic scientists and physicians. In 1921, Banting and Best conducted a number of experiments at the University of Toronto to demonstrate that removing the pancreas from a dog would induce diabetes. They subsequently removed fluid from a healthy dog's Islets of Langerhans and injected them into the diabetic dog to normalize its blood sugar. With the help of the biochemist J. B. Collip, they then extracted a pure form of insulin from cattle and succeeded in using this to treat the first human diabetic patient a year later in a Toronto hospital. Within another year, insulin was made widely available to patients. So this was far from the traditional notion of a long, continuous, and linear R&D process with discrete phases of development.

Indeed, belief that there might be a problem at the laboratory-clinic interface is not itself new. It was mooted in the early 1970s by Woolf (1974) in a *New England Journal of Medicine* editorial entitled *The Real Gap between Bench and Bedside*. So the leitmotif of linearity in retrospective accounts of basic and applied science does not necessarily correspond to actual scientific practice, which begs the question of what is new in contemporary accounts of the health innovation challenge. In her historiography of biomedicine, Lowy (2011) writes:

> World War II is usually presented as a turning point in the "biomedicalization" process. It accelerated and intensified collaboration between biologists, clinicians and industrialists, a development exemplified by the wartime production of penicillin. In industrialized countries, the post-World War II era was also characterized by important increases in public funding for medical research, the extension of health insurance to large parts of the population... and the rapid growth of the pharmaceutical industry. Of course, the separation between pre- and post-World War II circumstances is not absolute: laboratory sciences were intertwined with clinical practices from the early twentieth century. (Lowy, 2011: 117)

I explore this issue of novelty, which Lowy's quotation undermines, and interdisciplinary R&D practices in much more detail in the following chapter.

Although health innovation is not the crude linear process that is often portrayed—as I have outlined above the basic and applied sciences are not temporally and geographically distinct as often assumed—the linear model is still routinely used to frame biomedical R&D policy and practice. The linear stages of therapeutic R&D are perhaps more an artifact of the regulatory system, which demands the presentation of research in distinct, sequential phases. This elides the parallel processes and heterogeneous actors and innovation networks that actually shape R&D, as described in detail by Hara (2003).

However, the concept of linearity, which is easily reified by a casual and uncritical rhetoric of translational gaps in drug "pipelines," still continues to drive R&D policy and management (Tait and Williams, 1999; Williams, 2006).[6] One interview respondent captured this problem with the linear model when he stated:

> I think one of the challenges in this whole area is that the linear model of drug development is overly simplistic and, whilst it had enormous strengths in persuading those in the Treasury [United Kingdom] as to where the gaps might be, in the real world scientific discovery or even therapeutic development is in no way as simple as that. (Director of Policy 1, Public Sector Organization)

Here, linearity has strategic value in securing political support and funding for particular types of science. The tension between linear models of innovation, bench to bedside relations, and the novelty of TM will become more apparent later when I discuss practitioner accounts of its role and scope. For now, I simply want to emphasize that discussions about basic and applied research, assumptions of linearity in R&D, and expectations that "translation" might fix the problem of the broken middle in health innovation are closely connected. They also foreground the multiplicity of framings and meanings ascribed to TM and shape institutional and organizational practices and values within the health bioeconomy. I discuss the latter in chapter 4.

The Foundations of the "Broken Middle"

Arguments proclaiming that there are translational gaps in the health innovation system tend to focus on specific hurdles and constraints along the conventionally understood "bench-to-bedside continuum." Again, the language of a continuum unhelpfully creates the illusion of strict linearity. Hurdles that are routinely highlighted include not only cultural, institutional, and economic barriers that inhibit suc-cessful translation of discovery science into viable clinical products, but also more tangible challenges facing drug developers. These include lack of sufficient efficacy and safety in phase 2 clinical stud-ies, onerous and costly regulatory systems, rising R&D costs, patent expiry on blockbuster drugs with few products to replace them, and the organizational challenge of moving from small-molecule drug development to novel therapies based on new biology.

To restate the arguments of chapter 2, mature product pipelines and the difficulty of identifying viable business models for novel life science therapies have contributed to industry's anxiety about R&D and the long-term sustainability of blockbuster drug development. Although there is debate about the nature and extent of "innovation deficit" in the pharmaceutical industry—some authors ask if declining innovation is actually a myth (Schmid and Smith, 2005), and others maintain that reduction in R&D productivity is the result of a concentration of R&D efforts in high-risk research for unmet medical need, rather than a lack of innovation (Pamolli et al., 2011)—companies believe they are no longer producing enough high-value therapies to sustain growth. Furthermore, according to the bibliometric analysis conducted by Rafols et al. (2014), productivity from in-house R&D is continuing to fall. The so-called biotechnology revolution has yet to prove the panacea for industry woes and bring about a truly revolutionary era of therapeutics based on new biology (Hopkins et al., 2007; Kraft, 2004). The number of new drug approvals also continues to decline, despite increasing year-on-year investment in R&D (Kaitin, 2010). Furthermore, less than 25 percent of promising biomedical discoveries result in published clinical trials, and less than 10 percent become established in clinical practice within 20 years (Drolet and Lorenzi, 2010). It is therefore no surprise that many industry stakeholders believe there is a fundamental problem with the prevailing blockbuster model of drug development.

However, the notion of a broken drug innovation system has also been widely discussed and embraced in a number of key reports by scientific organizations (Academy of Medical Sciences, 2011; Cooksey, 2006); regulatory agencies (FDA, 2006); and funders of medical research (MRC, 2008; NIH, 2010, 2011). This reflects the diverse set of interests and systemic issues now at stake. The broken middle of health innovation is not merely a concern of the multinational pharmaceutical industry as it strives to maximize economic returns on therapeutic innovation and create efficiencies along the whole innovation pathway. The public sector has been coopted to share in this generalized anxiety about R&D productivity and contribute to finding a solution in the name of public health. There is a consensus across these reports by public sector bodies that there are entrenched problems in the middle stages of R&D, which require public support for greater translational activities, and the development and uptake of new tools to enhance drug discovery, development, and regulatory processes. A key issue, which has become a canonical theme in much of the TM literature, is the identification and validation of biomarkers to facilitate drug development and delivery, which I discuss in more detail later. Central to many of these discourses on TM are imagined

futures in which the exploitation of technologies within a new organizational framework contributes to solving the current challenge of
a broken R&D system. This, it is hoped, will deliver the benefits of
improved therapies and economic return on innovation to industry,
patients, and broader society. There are a number of drivers and strategic priorities for TM being put into practice by different constituencies, each with their own expectations and notional ideas of value and
benefit. These should not, as many authors now agree, be simply dis-
counted as hype (Brown et al., 2000; Morrison and Cornips, 2012).
They have material consequences for R&D and innovation.

Industry, Academic, and Policy Drivers

As both a general overarching philosophy and set of concrete practical
activities, TM has acquired increasing status in academic medicine,
the biopharmaceutical industries, and policy/regulatory communities
as a means of capitalizing on life science investments and contributing to the knowledge-based bioeconomy (OECD, 2009). It is also
expected to provide tangible benefits in terms of safe and effective
therapies for unmet medical need. If we unpack these narratives, we
can observe different notions of present and future value being mobilized across a variety of scientific, clinical, commercial, and political landscapes. Despite enormous investment in the field, there is, as
we shall see later, little consensus on the definitional and conceptual
boundaries of TM, its role in clinical practice, and what it can realistically deliver in terms of economic and social/clinical value. This is
partly a result of different practitioners using the term in a variety of
institutional and professional contexts, so that a simple, unified vision
of its aims and objectives is yet to emerge. This lack of consensus may
become a problem if incompatible visions and expectations (Borup
et al., 2006) become entangled within emerging institutional and
organizational structures. There are three key constituencies pushing
this broad TM agenda and each is a fundamental component of the
health innovation ecosystem and defines its multiple value chains and
orders of worth.

For the pharmaceutical industry concerned about phase 2 attrition rates, TM has acquired several meanings and driven a range of
organizational and management strategies. There are now TM units
within most major pharmaceutical firms. In some firms, TM groups
facilitate direct connection between basic research and patient care to
address key questions about how therapies will work in the clinical
setting. This has become known as the "patient-centred approach"

and includes attempts to perform first-in-human studies much earlier in the development process, as described in detail by Milne (2013). So here we can see the emergence of an explicit recognition of patient value as a driver of innovation.

Firms have also tried to bridge the gap between late discovery and early clinical development in an attempt to "de-risk" candidate drug selection and improve decision making on what products to take forward into clinical trials. The idea here is to learn as much about the clinical effects of a product very early in development to increase the chances of selecting a drug candidate likely to be successful. TM units have also served as conduits for valuing and accessing new external knowledge, technologies, and expertise through collaborations (Mittra, 2007). Industry is experimenting with various TM initiatives to improve the *business* of drug development and respond to the pressures being placed on blockbuster drug discovery, which I explored in the previous chapter.

Academic science and clinical medicine are supporting TM ostensibly to exploit the range of new technologies emerging from life sciences. They also wish to encourage communication and sharing of knowledge and expertise between the bench and the bedside, which many believe have become too institutionally and culturally distinct. Scientists and clinicians believe that a gulf has emerged as a consequence of increasing specialization on both sides, so TM is embraced as a mechanism to better coordinate and integrate research and clinical activities. Resurgent interest in the role of the "clinician-scientist" (a professional equally adept at working in the laboratory or the clinic) is indicative of this broader concern about the laboratory-clinic interface (Wilson-Kovacs and Hauskeller, 2012). The academic and clinical sectors have built hopes and expectations around a particular vision of TM that they hope will lead to step-change improvements in knowledge and understanding of key mechanisms of disease and diagnostic procedures (Academy of Medical Sciences, 2011). So value here is defined in terms of contribution to underlying science and clinical practice. This is in contrast with the more narrowly focused commercial expectations of industry, which are very much tied to the economic bottom line and captured in the rhetoric of efficiency gains and the derisking of candidate compound or drug selection. A senior academic clinician described the academic drivers and interests, in contrast to the commercial ones, in terms of "practical problem-solving driven by scientific curiosity" (Senior Academic Clinician 1). The tensions between these two views of translation will become more apparent in the following chapter, when I focus on the transformation of R&D practices.

Finally, there is a complex assemblage of social, regulatory, and policy institutions and organizations embracing and promoting the TM agenda. They focus on the safety and cost-effectiveness of new drugs and are actively thinking through how best to get innovative, path-breaking therapies from the bench to the bedside. The policy and regulatory communities are seeking to: (1) facilitate innovation of novel therapies and improve standards of safety and efficacy, for instance, through the use of biomarkers and new diagnostic testing; (2) improve the design and execution of clinical trials by utilizing improved preclinical knowledge; and (3) contribute to the growing bioeconomy through investment in new innovative technologies, therapies, and small companies. Consequently, there has been substantial government and charitable investment in translational R&D in universities and other public sector organizations; particularly from the Medical Research Council (MRC) in the United Kingdom and the NIH in the United States. I now explore in some detail the NIH's translational strategy and its expectations of clinical value from TM, and also reflect on similar initiatives that have developed in Europe.

The NIH as an Anticipatory and Promissory Organization[7]

For most of its history, the NIH has focused its funding and research priorities on basic research.[8] Crowley and Gusella (2009) argue that this was a direct consequence of Roosevelt's postwar R&D policy built on Vannevar Bush's vision and support (formed during his experiences working on the Manhattan Project) for strong basic science—encapsulated in his famous essay *Science: the Endless Frontier*:

> The Bush model has had profound implications for science policy, the organization of biomedical research communities, and science funding both locally and globally...the model logically suggested that medical schools and academic health centers (AHCs) should preferentially recruit basic scientists and that their careers be well supported by generous allocations of research space, facilitated promotions, and prestige. An unintended consequence of the Bush model was that human research became relegated to a far downstream component of the scientific discovery process, essentially serving as little more than proof of a given scientific principle, rather than as a means of defining new science or a real opportunity to influence health care. (Crowley and Gusella, 2009: 1)

The influence of this model was reflected in the NIH funding twice as much basic research as clinical research. It also had a direct role in changing the culture of medical schools and AHCs to prioritize very early stage research. Translation of basic science into viable clinical therapies became the sole responsibility of the pharmaceutical industry. However, over the past couple of decades there has been a recognizable and significant shift in emphasis from basic science to applied clinical research and translation. I argue that the NIH has become an anticipatory and promissory organization (Pollock and Williams, 2010), which, through its recent policies and strategies, has created and shaped expectations about technology futures and driven particular therapy options in biomedicine. It is also helping shape the long-term value of new biology. Indeed, strategic decisions made by organizations like the NIH do enact particular values and valuation practices that have a material impact on the innovation ecosystem and the practices therein.

The changing priorities of the NIH are perhaps best captured by the terse statement that the current director of the NIH, Francis Collins, made in 2009 to the *New York Times*, when he said: "We're not the National Institutes of Basic Sciences...We're the National Institutes of Health" (Francis Collins, cited in New York Times, October 5th, 2009). However, it was the specific set of initiatives set up by the previous director of the NIH, Dr Elias Zerhouni, between 2002 and 2006, which signaled the NIH's commitment to TM and the breaking down of perceived barriers between basic and applied medical research.

The NIH's[9] 575 million USD investment in the National Center for Advancing Translational Sciences (NCATS), which had a remit to catalyze innovations in translational science and to improve innovation in drugs, diagnostics, and devices, is perhaps most indicative of the science and policy community's growing commitment to the field of TM (NIH, 2011). This new Center, which was established in 2011, represented a significant amount of public funding dedicated to the challenge of bridging the supposed gap between basic science and clinical application. The NCATS brought together, under one national center, a number of translational initiatives that the NIH had been implementing for a number of years.

Before the emergence of the NCATS, translational activities were being developed and implemented in many different ways across a variety of specific NIH funding schemes and centers. In defining TM, the NIH points to two key areas. First is simply the application of discoveries made in the laboratory and preclinical research to the clinic through the support of human clinical trials. Second is the promotion of best-practice health care within the community, which is a much broader definition and is rooted in the notion of health services research and dissemination. This is in the realm of "T2" research, which I explain in more detail later. Woolf (2008) points out that this second aspect of translation within the NIH is not as well funded nor given the same strategic priority as the first. The NIH also considers cost-effectiveness of preventing and treating disease as an important facet of the translational research agenda. For the NIH, TM is envisioned as a journey through which the outputs of basic laboratory science are transformed into tangible health benefits, through a series of translational steps that require direct institutional support and guidance.

The NIH Roadmap and Common Fund

The "NIH Roadmap for Medical Research" was the result of a major NIH consultation, set up by Elias Zerhouni in 2002 to explore the challenges facing medical research in the United States. It reflected the growing emphasis on translation as a key public policy issue. The Roadmap established a list of priorities and strategies to address the challenges of converting basic science into viable therapies to benefit patients (Zerhouni, 2003). Kraft (2013) argues that part of the impetus for the Roadmap was to ensure insights from the HGP would inform clinical practice. This was very much about capitalizing on the high public investment that went into the HGP:

> Zerhouni defended the *Roadmap* with a rhetoric that emphasized the need to engage with the "new realities", which required more tools, cross-disciplinary teams, and an overhaul of the infrastructure for clinical trials. As he argued, it provided a way of "synergizing areas that no institute either has the mission or the resources to invest in." For him, the *Roadmap* was setting out a process to establish translational research as the new paradigm in biomedical research—which would foster the alignment of all constituencies within the innovation process, to develop "sustainable and integrated efforts in translational and clinical research that can yield new products, approaches, and diagnostic tools in an efficient seamless, manner." (Kraft, 2013: 43)

From Kraft's analysis, we can see the NIH strategy shifting away from basic research, which it historically prioritized, to more downstream drug development that had for many decades been the preserve of the large pharmaceutical firms. While some traditionalists decried the significant shift in strategic priority, this was perhaps a growing sign of the times for publicly funded biomedical research.

The public sector vision for the future of therapy development was one in which it had a much clearer and central role in the development of new drugs.

A set of initiatives were launched in late 2004 as five- to ten-year high-impact programs to improve the health innovation cycle and remove the major roadblocks identified in the Roadmap. Initially, initiatives were funded by small contributions from each of the NIH's 27 institutes and centers (ICs). This changed in 2007 with the creation of the NIH Common Fund, which was responsible for delivering on the Roadmap initiatives. In 2010, the Common Fund budget was 544 million USD, which at the time represented 1.8 percent of the NIH's total actual obligations (NIH Office of Budget, 2011).

There are 25 major roadmap initiatives financed by the Common Fund. They have a strong translational element.[10] Of particular importance are the exported programs, the Clinical and Translational Science Awards (CTSA) and the Clinical Research Policy Analysis and Coordination (CRpac). The first provided funding and support for a consortium of clinical and translational research centers throughout the United States. The second served as a focal point for the coordination, streamlining, and optimization of policies and requirements for the successful conduct and oversight of clinical research. In addition to these two programs, there is a regulatory science initiative, which is a partnership with the FDA, to improve tools and processes for evaluating novel therapies. For now, I wish to focus on the CTSA program, as it represented an explicit policy to respond to the broken middle problem and anticipated TM as a high-value area for investment to improve and refine future health research.

The CTSA Program

The CTSA program was established in 2006 and was originally funded and governed through the National Center for Research Resources, but now falls under the remit of the new NCATS. The mission of the CTSA is to transform at a local, regional, and national level the environment for biomedical innovation and enable more efficient translation of basic science knowledge into effective treatments that will improve human health (National Center for Research Resources, 2009). More recently, the CTSA website includes reference to community-engaged research to emphasize the full spectrum of translation from bench to bedside. Approximately 60 academic institutions in the United States have received CTSA funding, totaling around 500 million USD. As of 2012, the CTSA program was the largest dedicated clinical and translational research program in US history (Califf and Berglund, 2010). So this was not simply piecemeal funding of a "trendy" new approach, but represented a real public commitment in terms of both capital and resource.

The CTSA has five key strategic goals: (1) to build national clinical and translational research capability; (2) to provide training and improve career development of clinical and translational scientists; (3) to enhance consortium-wide collaboration and partnerships to build new translational networks; (4) to improve community and national health; and (5) to advance T1 translational research and move basic laboratory discoveries and knowledge into the clinic (Clinical and Translational Science Awards, 2011). At the heart of this initiative is the belief that the diversity, size, and scope of the institutions involved in the CTSA will enhance the program's overall impact. CTSA institutions are expected to promote and sustain clinical and translational science by providing a conducive environment, including sufficient infrastructure, funding and training, for researchers in this interdisciplinary field.

Central to the vision of the CTSA has been collaboration and communication, and interdisciplinarity is now a central theme in much of the NIH literature, which I discuss in more detail in the following chapter. However, one of the key elements in the CTSA program, which speaks to a broader trend in public involvement in biomedical research, was the promotion of public-private collaboration, in addition to the building of broader university researcher networks. The CTSA consortium has a dedicated Public-Private Partnerships Key Function Committee to improve links between CTSA institutions and commercial organizations (Clinical and Translational Science Awards, 2011). Furthermore, there is a "CTSA-IP" web tool that collates all the technologies developed by CTSA institutions and the NIH, enabling commercial companies to identify valuable partnership opportunities. In February 2010, the CTSA hosted an industry forum event, which included representatives from the largest pharmaceutical companies, to explore how the public and private sectors might collaborate more effectively.

Overall, the CTSA program and other NIH initiatives have, for many years, aimed to overcome the various barriers and challenges facing contemporary health innovation. The recognition of a need to invest in a broader set of interdisciplinary and cross-sectoral biomedical research areas, and particularly the more downstream and applied phases of health R&D, is novel and represents a real change in public sector science strategy. Indeed, one could cynically argue that TM has been pushed by the policy community as a means to channel public funds to downstream drug development processes without having to explicitly state this aim, which would not always be politically expedient. This raises the question of whether public money should be invested in the financially more risky drug development work that has traditionally been borne by the commercial pharmaceutical industry. Also, where does long-term value reside for the public funders as opposed to the commercial ones? I explore this in more detail when discussing public-private partnerships in the following chapter, where I argue that we need to employ a much broader concept of value to fully understand and provide justification for these kinds of public investments.

Translational Initiatives in Europe

The NIH translational strategies and underlying philosophy have been implemented across the globe in a variety of different contexts. In a similar vein, but on a much smaller scale, the United Kingdom's' MRC launched six translational medicine centers with funding of £15.5 million in 2007, which aimed to develop research programs with clear milestones to overcome existing gaps or hurdles in translational science (MRC, 2007). The MRC's translational research strategy, as outlined in its Strategic Plan 2014–2019 (MRC, 2013), described how translation is important for both improvement in human health and economic benefit. Once again, the needs of the bioeconomy are an intrinsic feature of early-stage funding strategies for biomedical research, with both economic and patient value intimately connected. The report also addressed the MRC's developing strategy for engaging with industry, through the establishment of the Biomedical Catalyst, in collaboration with the Technology Strategy Board (TSB), which is now known as Innovate UK. This is very much a translational organization with 180 million GBP to, according to the MRC,

> speed up the development of new interventions, and the initiation of completely new ways of working with companies: the development of disease-specific consortia to investigate why patients respond differently to drugs and a joint initiative with AstraZeneca making available clinical compounds to better understand diseases. (MRC, 2013: 4)

In future, the MRC has signaled its intent to further promote new treatments, diagnostics, and preventative strategies, and encourage inward investment and industry partnerships so that the United Kingdom remains at the forefront of biomedical research.

Innovate UK, in addition to collaborating and jointly funding initiatives like the Biomedical Catalyst, has been given responsibility for investing money into academic and commercial research that promises real commercial potential, in an attempt to overcome translational hurdles. A particularly illustrative example has been its funding of regenerative medicine, which I talk about in more detail later in chapter 5. The European Innovative Medicines Initiative (IMI) is another example of large-scale public sector investment in translational research, in this case an attempt to promote industry-academic collaborations to improve safety and efficacy of new and existing medicines and build large precompetitive research consortia.

The translational policy agenda has also been taken up by the United Kingdom's National Health Service (NHS). The National Institute for Health Research (NIHR), which was set up in 2006 to centralize and focus patient-centered research in the United Kingdom, created 12 Biomedical Research Centres (BRCs) with NHS and University partnerships to improve translation of health innovations into NHS practice. Snape et al. (2008) write:

> The main concept of a BRC is to create a structure that successfully brings together scientific investigators and clinicians committed to the process of TR (Translational Research). This process needs to be transformative in nature in order to surmount the significant barriers that prevent high-quality patient-based research. Specifically, facilities for patient-based experimental studies should be available adjacent to biomedical scientists and clinicians involved in TR. (Snape et al., 2008: 903)

This approach is expected to lead to direct and tangible benefits for patient care. The BRC initiative, alongside those set up by organizations like Innovate UK, was partly borne from the outcomes of the Cooksey Review (2006), which cited an unsupportive culture in the NHS, institutional barriers, and perverse incentives—such as greater priority and reward structures for basic research than for applied research—as presenting barriers to good clinical outcomes and patient value. Coupled with a lack of coordination in funding, the Cooksey review identified major barriers to translation that needed a formal policy response, and this drove many of these large-scale translational activities.

All these initiatives are indicative of a broadening policy drive to invest in new kinds of approaches to biomedical research and therapy development, and to contribute to the growth and management of the health bioeconomy. Each reveals certain expectations about the future value of biomedical research and therapy. These initiatives also anticipate a future when the broken middle of health innovation can be overcome and the health and wealth benefits of new biology realized within both the public and commercial sectors. In this context, they are not simply anticipatory organizations, but also promissory organizations that are proactively shaping the nature and future prospects of the innovation ecosystem.

A Generic Model for Translational Medicine?

That there have been different imperatives driving TM, and a variety of strategies for implementing it, reflects the range of different sectors' needs, expectations, and anticipation of value. It also reflects their particular understanding and framing of the core problem for which TM is envisioned to be the most appropriate solution. Nevertheless, there does appear to be consensus that a systemic problem exists in the middle stages of health R&D, which requires new approaches in terms of how science, technology, infrastructure, and resources are organized and managed.

A basic lexicon for TM has emerged in the literature, with one popular model expounding three distinct phases of translation (Dougherty and Conway, 2008). "T1" refers to the translation of basic science into clinical efficacy and is focused on the early stages of drug discovery and preclinical testing. "T2" refers to efficacy translated to clinical effectiveness, and focuses on the middle stages of R&D. These two reflect the translational priorities of organizations like the NIH. "T3" refers to effectiveness translated to healthcare delivery, so is very much rooted in late-stage development. Drolet and Lorenzi (2010) take this approach further by distinguishing a "zone of translation," which is an intermediary between basic science and accepted clinical practice/overall societal health impact. For these authors, T1–T3 represent particular chasms in research progression along the bench-to-bedside continuum, and translational research refers to those specific activities whose purpose is to bridge these chasms. Although this is still quite a linear description of TM, it usefully highlights some of the different sites and interstices of knowledge where practitioners believe better translational mechanisms are needed.[11] It is with this general TM framework in mind, and understanding of what is driving the approach, that we can begin to explore practitioner narratives, understandings, and values around TM.

Practitioner Understandings of Translational Medicine

In this section, I analyze the underlying definitions and framings of TM from the perspective of different TM practitioners, before exploring a specific and crucial focus of TM, namely the identification and use of biomarkers. The value currently ascribed to biomarkers, and high expectations about their role in mitigating phase 2 attrition rates, highlights the social, political, and economic salience of the broken middle argument and its impact on biomedical research.

The definition of TM and its conceptual and practical boundaries is a topic of much debate within the biomedical science and policy communities. The T1–T3 model simplifies what are quite complex and diverse beliefs about the R&D challenges and appropriate strategies for responding to them. On definitions and boundaries, a number of views emerge from the scientific literature and interview accounts of key professionals, which map on to one or more aspects of the T1–T3 model. Definitions range from the specific to the general and can cover "organizational processes" as well as "scientific application." One interview respondent, from a major funding organization, argued that translational *research* is not area of science but a process of bidirectional knowledge flow from fundamental research to application and back again. Note again here the implicit assumption of a distinction between basic and applied science, which is temporally distinct. The respondent proceeded to suggest that translational *medicine*, in contrast, is a subset of research focused on what has traditionally been called "experimental medicine." It is interesting to note that the definition of experimental medicine, like TM, is also open to some debate. According to this respondent, some believe it should encompass epidemiology while others think it should be limited only to small patient studies. Nevertheless, terms like TM and translational research are used interchangeably by practitioners, and different framings may cover organizational or institutional processes as well as specific applications of science and technology. So we can now identify both narrow and broad definitions and framings of TM.

Narrow Definitions/Framings of TM

Narrow framings of TM tend to draw on the concept of "applied basic science" and often emphasize the role of new biology and life science technologies. For example, TM has been described as a process for determining treatment based on molecular biological characteristics (Saijo, 2002) or as the "translation of genomic and functional biology discoveries into clinical practice" (Niederhuber, 2010: 1088). TM is often reduced to a discrete set of genomics-based techniques and applications, which can be used as a conduit for integrating different types of knowledge and expertise at the bench and the bedside.

One interviewee stated: "I think that the definition currently of translational medicine is probably DNA-based or protein-based type of biomarker studies" (Senior Academic Scientist 5). Another senior academic emphasized the benefits of a narrow definition when he stated:

> I think in many ways translational medicine is a very murky term...I think that a narrower definition gives some clear goals and directives and ways of unifying the academic and industrial community in partnership. (Senior Academic Scientist 4)

Here, the importance of clearly defined outcomes from academic-industry collaborations is presented as a key feature in setting the boundaries of TM. This is very much an output or goal-oriented view of TM, where its underlying value rests on an ability to generate collaborative and intersectoral research partnerships.

We can also observe some sector-specific framings of TM, with the pharmaceutical industry, for instance, reducing TM to the process of commercial drug development. An interview respondent who heads a commercial clinical trial company gave the following account, which is very much rooted in a commercial bench-to-bedside notion of TM:

> What we're looking at is taking something that perhaps is defined at bench level in terms of a particular drug or something that targets a particular site and then that is developed through a whole range of processes to the point where it can be accepted as a potential drug target to work on through a pharmaceutical company, and then eventually into the clinical side. So the way that we would define TM is taking something that is very much research-oriented and translating that into a commercial product. (CEO of a Clinical Trial Company)

Similarly, another respondent stated that TM was simply

> translating experimental findings in the laboratory through to clinical findings in the hospital setting...we're trying to develop drugs to treat established diseases and we need to predict what might happen in the clinic. (Senior Business Manager, Large Pharmaceutical Company 6)

Responses from industry suggest they adopt a process-driven definition of TM with a clear commercial focus on improving efficiency of R&D and reducing phase 2 attrition. This narrative is rooted in the notion of a bench-to-bedside continuum, which assumes that in the middle stages of a sequential R&D process there is a fundamental problem that needs to be fixed.

In contrast to this commercial view of translation, clinicians tended to frame TM predominantly in terms of using life science technologies to improve diagnosis and categorization of disease. One Professor of Clinical Psychiatry stated:

> The studies we have done to identify genes in schizophrenia, bipolar disorder, and depression can all be considered highly translational because they are aimed at identifying sub-populations of psychiatric diagnoses to improve treatment studies At present in psychiatry we are quite good at defining [these conditions] with operational criteria which are reliable and any two psychiatrists will agree most of the time on the diagnosis. However, the biological validity of these diagnoses are unknown and untested and everybody accepts that our [current] diagnostic categories don't have any real biological validity. If genetic studies lead to clearer diagnoses in psychiatry this will translate into better treatment studies. (Professor of Clinical Psychiatry)

This account presents TM as a valuable mechanism for exploiting life sciences-based tools and technologies to better categorize clinical disorders and ultimately improve patient treatment, which was a recurrent theme in the accounts of both academic scientists and clinicians.

These narrow framings of TM prioritize the scientific, technological, and clinical processes of TM, rather than the broader institutional and system-level dynamics that are perhaps more relevant to the implementation and exploitation of new organizational models. They also appear to reify the bench-to-bedside continuum (with the conventional demarcation of basic and applied research) in the spirit of the T1–T3 model. There is little or no emphasis on feedback loops from the clinic to the laboratory, or on the parallel processes that can often be temporally and geographically disjointed in modern health innovation systems.

Broader Framings and Emphasis on the "Bench-to-Bedside and Back Again" Process

Some authors consider the one directional bench-to-bedside approach to TM outdated and unhelpful. Instead, they define TM as a two-way iterative process from bench to bedside and back again. Here, knowledge, information, and expertise are continually shared between clinicians and laboratory scientists so that patient data can explicitly inform basic science (Ledford, 2008; Soderquest and Lord, 2010). Mankoff et al. (2004) have argued that the unidirectional definition fails because animal and other experimental models are not truly representative of human pathology. Many interview respondents argued that a feedback loop from bedside to bench is crucial. An industry respondent stated:

> First of all it goes both ways, because a lot of the stuff that we have discovered from doing this in humans [treating with experimental drug] was then translated back into the lab. It's not uni-directional. (Senior Scientist, Large Pharmaceutical Company 7)

Similarly, a respondent from the policy community argued that this way of framing TM takes us away from crude, linear accounts of R&D:

> The translational medicine element that I think is really beneficial is the fact that, rather than it being a linear process, there is this two-way feedback...the science is definitely being influenced by patient accessibility in application. (Senior Policy Manager for Government Agency 2)

Rubio et al. (2010) have developed a broad working definition of TM, which emphasizes multidirectional integration of basic research, patient-oriented research, and population-based research with the long-term objective of improving public health. The NIH (2010), while tending to prioritize T1 and T2 translation, as discussed earlier, also describes research that facilitates the use of best practice health care within the community, and ensures cost-effective treatment of disease, is recognized as an important component of TM. This goes slightly beyond the T3 phase of translation described earlier and provides a fuller and more systemic account of translation. As one academic scientist stated:

> My understanding of translational medicine is converting fundamental biomedical discoveries into practical solutions for health problems. Mostly it's in the form of drugs, but it's also in terms of policy and other things. So, the discovery that smoking is bad for your health was a major translational achievement where somebody's fundamental epidemiological studies followed up by some animal experimental studies clearly indicated that smoking was bad for your health, and was perhaps the major component of lung cancer. And that's been reinforced over the years and given rise to policy change, which has given rise to measurable benefit. That's an example of translation in the policy field. (Senior Academic Scientist 6)

This account exemplifies the broader institutional policy dimension and presents TM as being concerned with much more than the conventional drug development pipeline model. It also alludes to the important role of formal legislation (such as bans on smoking in public places) as a nonmedical translational public health intervention. Similarly, Ogilvie et al. (2009), in their critique of simplistic, narrowly defined, and linear pathway-based models of TM, emphasize the important role of public health research and epidemiological studies. Through exemplar case studies, the authors demonstrate the limitations of conventional wisdom on how new knowledge becomes translated into practice. They show that the "public realm," broadly defined, is an important part of practice and illustrates the complex, nonlinear, and nonobvious influences on clinical science. One case example discussed by the authors is the epidemiological research that supported the "Back to Sleep" campaigns to change recommended sleeping positions to reduce the risk of Sudden Infant Death Syndrome (SIDS). The nonclinical observational study, according to the authors, was the only method capable of demonstrating effectiveness and could not be explained by a linear, TM approach (Ogilvie et al., 2009).

This range of views suggests that TM should be characterized as a general organizing principle, or "social technology" in Nelson and Sampat's (2002) terminology. From this perspective, it is an institutional mechanism for coordinating multiple professions, knowledge domains, economic/scientific activities, and ultimately different values within new and emerging organizational contexts. Ideas of TM drive the development of new mechanisms and processes to better bridge or integrate basic and clinical science and facilitate knowledge and information transfer from bench to bedside and, crucially, *back again*. However, even if we consider TM in this broadest sense, the question still arises as to whether there is anything novel in the current practices based on the "bench-to-bedside and back again" philosophy, and whether the very distinction between basic and applied phases of R&D adequately reflects the messy realities of contemporary life science innovation and the challenges that lie therein.

The "Novelty" of TM Practices in the Context of the Bench-to-Bedside "Problem"

As I discussed earlier, the conventional view of basic and applied science and the caricatured accounts of bench-to-bedside relations that are used to present TM as novel and cutting edge are often based on a misrepresentation of the history of clinical medicine and the professional and institutional boundaries between laboratory and clinic.

A number of authors have nicely illustrated this with case examples (van den Hoonaard, 2009; Martin et al., 2008; Stokes, 1997; Sturdy 2012). Sturdy, for example, argues that the tensions and conflicts between clinicians and bench scientists are often overstated in historical accounts, which implies that the problem of a broken middle has also been overstated and provides only a partial account of the health innovation challenge. Indeed, many of the barriers to successful health innovation, particularly in the context of novel innovations based on new biology, are broad and systemic. They include the impact of regulation, markets, and clinical uptake on innovation strategies (Mastroeni et al., 2012; Mittra, 2008; Tait, 2007), which are the focus of chapter 5. From this perspective, a TM approach focused only on a particular set of technology and knowledge integration problems in the middle stages of drug development will not be sufficient for improving overall health innovation.

Interestingly, a number of interview respondents had quite critical and nuanced perspectives on the putative novelty of TM. For example, most respondents agreed there has been a rebranding of conventional scientific and clinical practice, in the drive to secure research funding from a policy community that has become enamoured by the rhetoric of "translation," or has at least been using it to justify public investment in drug development. Some respondents expressed concern that TM is defined so broadly that it can cover almost anything vaguely related to applied basic science. An interview respondent from the United Kingdom's NHS, for example, stated: "They're buzz words; I used to call it applied research" (Health Service R&D Manager 2). A senior academic scientist stated:

> I'm not so sure it's novel because there have always been people pursuing translational research. Really what it's reflective of is an effort to brand something and use that brand to catalyse the movement of discoveries of basic research into clinical practice. (Senior Academic Scientist 4)

Other respondents agreed that the concept was not really capturing any inherently novel practices. Nevertheless, they believed it still served an important function in alerting the biomedical community to the significance of the R&D challenges, and the need to think of new ways to resolve them at the scientific, clinical, technological, and policy levels. It is important to recognize that those who see TM largely as a rebranding exercise do not deny that the conventional innovation pathway is broken and in need of fundamental repair.

Despite competing views about the role and scope of TM, and the fact that many of the practices underpinning TM are not novel in and of themselves, there has clearly been growing interest in efforts to influence bench-to-bedside relations, and much institutional resource and infrastructure (both public and commercial) dedicated to fixing what is considered to be a broken middle of health R&D. In recent years, biomarkers have become an emblematic feature of this growing TM agenda. Biomarkers have become the subject of powerful promissory discourses, but their value has been contested and expectations have been mixed. I now critically explore the nature and role of biomarkers in the health bioeconomy, and consider how their prioritization reflects certain assumptions about the current health innovation challenge.

The Promissory Value of Biomarkers

A biomarker is any objective, measurable indicator of a biological state or process. The value of a biomarker is linked to its ability to facilitate understanding of disease mechanisms/pathways or therapeutic safety and efficacy. Some conventional biomarkers are relatively simple and well established within the clinic, such as cholesterol as a biomarker for risk of coronary heart disease, or blood pressure as a biomarker for hypertension, but a number of novel molecular biomarkers have been identified since the mapping of the human genome.[12] Although there has been a particular focus on complex molecular or biochemical biomarkers, there has also been significant innovation in noninvasive imaging biomarkers, such as anatomical and functional imaging (Weber, 2006)[13]. Nussenblatt and Marincola (2013) argue that biomarkers have a variety of functional benefits, which include the potential to inform patient stratification for appropriate treatment (so-called stratified medicine), identify new targets for therapy, measure effects of treatment, verify hypotheses about mechanisms of drug action, categorize disease status, and provide surrogate markers for determining long-term benefit of treatment. Many different kinds of biomarkers are now being identified, and translational studies are trying to validate new biomarkers that will improve knowledge and understanding of disease, clinical decision making, and drug development processes.

Biomarkers have become a central theme in discussions about the role and long-term value of TM. They are treated as almost synonymous with TM, because they are considered relevant both to industry attempts to reduce phase 2 attrition rates and to academic and clinical understanding of disease mechanisms and patient outcomes. Some practitioners believe that biomarkers provide concrete foci for cross-sector and interdisciplinary TM collaborations. Such focused collaborations are perhaps easier to manage than projects built around more ephemeral areas of translational science, a view supported in the following account from a clinician:

> I think it's [biomarkers] a critical area certainly, and it's one that universities and medical schools can get engaged with relatively simply. Whereas the late-phase clinical trials are much more difficult for us to be engaged in. (Senior Clinician 2)

In general, senior academic scientists were also very optimistic about the potential role and significance of biomarkers. Some implicitly framed biomarkers as a progressive innovation that will come to replace many of the conventional clinical practices currently relied upon, as the following response nicely illustrates:

> Molecular biomarkers will replace a lot of conventional diagnostic tests, inevitably. Cancer at the moment is still defined, subdivided and graded by pathologists looking down microscopes...all of this diagnosis has to change in the next 20 years; it's still arcane. (Senior Academic Scientist 7)

Here, biomarkers are imbued with a great deal of promissory clinical value (or biovalue) that will enable scientists and clinicians to replace the old, subjective, and imprecise methods of the pregenomic era. Another interview respondent supported this view when he talked about biomarkers being of particular value to the field of psychiatry. He argued that current diagnostic methods for psychiatric disorders are inadequate, because there is often a constellation of complex symptoms with blurred boundaries and co-morbidities. The diagnostic categories themselves are largely based on the highly subjective judgments of the clinican. As Pickersgill notes, boundaries between psychiatric disorders are not always clear, leading to problems in diagnosis and treatment. For example, an enduring "lack of a consensus regarding the relationship between [antisocial personality disorder] and psychopathy...results in much uncertainty on the part of those who research and treat these disorders," with wide-ranging implications (Pickersgill, 2013: 335; see also Pickersgill, 2012, 2014). Molecular biomarkers, according to the clinical psychiatrist interviewed, could potentially provide a more robust and objective measurement of disease state and improve both diagnosis and treatment. However, this is still very much a future prospect or hope rather than a current practical reality.

In contrast to scientists and clinicians' primary interest in the value of biomarkers to improve classification and diagnosis of disease, the pharmaceutical industry's interest lies in their potential to identify safety or efficacy issues in the middle stages of R&D, so that cost of failure in phase 2 and phase 3 clinical trials can be reduced. Here, the value of biomarkers is inextricably linked to the notion of a broken middle of R&D, and the elusive quest for greater efficiency and cost savings. This industry narrative provides a more pragmatic and tightly defined role for biomarkers and has recently been used to justify industry participation and growing investment in various pre-competitive biomarker partnerships and consortia, including in the United States a major public-private Biomarkers Consortium, which is managed by the Foundation for the NIH.[14] I discuss this example in more detail in the following chapter.

There is additional interest in the potential for biomarker data to be used in regulatory decision making, both to provide surrogate endpoints for clinical trials and to select patients for clinical studies. This particular application is not being driven solely by the pharmaceutical industry. Regulators, such as the US FDA, have outlined a commitment to the identification and validation of biomarkers and innovative clinical trial design in order to drive forward pharmaceutical innovation (FDA, 2006). Similar approaches are being considered by the European Medicines Agency (EMA). The FDA has also established an initiative to facilitate the development of biomarkers and ensure that regulations for the associated diagnostic tests are fit for purpose (FDA, 2011), which is indicative of growing expectations for the technology. However, as Moreira at al. (2009) argue in the context of using biomarkers to identify preclinical dementia, the deployment of collective uncertainty is often a key feature of these kinds of initiatives, and the development of new standards and categories can often increase rather than reduce ambiguity. While it is clear that biomarkers are being embraced by a number of diverse actors and organizations involved in biomedical research and innovation, it is important to subject these promissory narratives and expectations to critical analysis, and also to consider the assumptions that this focus on biomarkers makes about the nature of the health innovation pathway.

Limitations of Biomarkers

The sociological literature on hopes and expectations tends to focus on how promises and expectations shape the early-stage development of new technologies. However, they also camouflage the risks, anxieties, and other social/cultural processes that may constrain technological innovation. This absence can be ascribed partly to the fact that negative expectations are often not articulated in the early phases of technological development. To attract the interest of relevant actors, early promises tend to paint a rather rosy future of the technology and often contain utopian, technology-driven dreams about how the world will look if everyone were to use the new technology. (Oudshoorn, 2011: 365–336)

Furthermore, as Williams (2006) points out in his work on "compressed foresight," expectations (either positive or negative) can help reify in the present a particular future trajectory that is then liable to technological "lock-in" and obdurate policies and strategies that are ultimately unhelpful. In the case of biomarkers, there have been critical voices urging caution about unsubstantiated claims regarding their short-term value. These represent the negative expectations that Oudshoorn rightly points out are an important element in the construction of technology futures. It is important to consider whether specific examples of TM, such as biomarker discovery and validation, are based on untested, and perhaps unrealistic, assumptions about the transformative impact they are likely to have on the short-term development of diagnostics and therapies. Are expectations around biomarkers an example of compressed foresight? Some interview respondents objected to what they saw as the fetishizing of biomarkers as a panacea for the broken innovation system. One academic scientist, specializing in hematology, described how biomarkers for indicating the fragility of plaque are considered the "Holy Grail" within his field. Researchers want to understand when plaque is about to rupture and have developed a number of techniques to try to identify this in real time. He claimed that they have attempted to measure markers in the blood stream, such as metalloproteinases that are shed from the plaque. They have also used imaging techniques to visualize the plaque and see if it "lights up" with a positron emission tomography (PET) ligand, which would be indicative of very active plaque and a potential target for a therapy. However, the respondent added the following crucial caveat:

These approaches sound quite mature and well thought out but what they don't address is the fact that there are hundreds of plaques in the average vasculature, and some of them are vulnerable and some of them not, and what you see really is an aggregation of all these things, and if you were trying to find the plaque that killed you, you'd be on a hiding to nothing. So it's helpful, but not as helpful as people would try to make out. (Senior Academic Scientist 2)

Incomplete knowledge about biomarkers, and a tendency to grant them special status in clinical decision making and/or commercial drug development programs, can also lead to false conclusions about process and outcomes. The respondent continued to state:

> Let me give you the example of estrogen. If you give it as hormone replacement therapy [trials have shown] it lowered LDL [Low Density Lipoprotein] cholesterol, it raised HDL [High Density Lipoprotein] cholesterol, it did a host of other things in the artery wall that you would have said, right, this is absolutely cast-iron, we're ok here, we'll get benefit...At the end of the day, estrogen caused more heart disease, and the biomarkers would have driven this in entirely the wrong direction. Because, what you cannot get from biomarkers is the aggregate effects...I think we're decades away from having enough biomarkers to understand the entirety of the process, and then aggregating them is very difficult. (Senior Academic Scientist 2)

The skepticism evinced in these two accounts run counter to the more optimistic and transformative views of the use of biomarkers in drug development and diagnosis. The underlying complexity of disease processes and treatment effects, from this more skeptical perspective, renders current biomarker studies insufficient as a replacement for conventional clinical studies and outcome measures. The promissory and sanguine vision of a biomarker-led drug development and therapy paradigm is therefore very much a projected future, one that must overcome current technological reality and clinical complexity, even though the rhetoric evokes a greater sense of immediacy or inevitability. This has implications for their more general and extended use in regulatory decision making for clinical trials and clinical practice, which continues to be anchored to the conservative and cautious "big pharma model" of drug development (Tait, 2007).

In their paper on gene mapping, Terwilliger and Goring (2009) provide a compelling argument that many future strategies around genomics have been made on unrealistic and untested assumptions about what the technology can realistically deliver. This argument would seem to apply equally to biomarkers and, perhaps, TM more generally. The diverse and sometimes contradictory views of interview respondents about the benefits, limitations, and ultimate value of biomarkers would suggest that these technologies may work in some fields, but not necessarily in others. For example, biomarkers might facilitate better characterization, diagnosis, and treatment of certain cancers, but may not be so helpful in other areas, such as diabetes for instance, where there might be 30 known markers that are only slightly associated with raised risk level. In the latter case, we must be skeptical about the real underlying power and subsequent clinical value of the technology.

Although biomarkers do promise a number of solutions to the purportedly broken conventional model of drug innovation, there is a clear danger in fetishizing one technological solution, and ignoring broader systemic challenges and constraints. Furthermore, this focus on biomarkers, which from an industry perspective is very much rooted in concerns about phase 2 attrition rates, does tend to assume a particular innovation pathway model for drug development, with a chasm between basic and applied research seen as responsible for a lack of successful translation from bench to bedside. TM then emerges as the most obvious and inevitable solution. However, I want to emphasize that the broken middle of R&D narrative is far more complex and contested, and the distinction between basic and applied research, or laboratory and clinic, is not as straightforward as is often presented.

Conclusion

In this chapter, I have suggested that TM is more than a discrete set of technological instruments and mechanisms for exploiting the life sciences for therapeutic benefit. It is also based on a number of shared assumptions about the nature of R&D and the current challenges of drug development, particularly phase 2 attrition and a perceived gap between the laboratory and the clinic. The data I have presented reveal the complex and relatively fluid definitional and conceptual boundaries that are employed by different professionals as they envision various objectives and outputs for the field. Driving these discursive narratives has been a particular set of perceived health innovation challenges, which I have referred to as the "broken middle" of health R&D. This problem narrative has presupposed a particular role and scope for TM, the centrality of biomarkers providing a key illustrative example.

What I have attempted to demonstrate in this chapter is that TM remains a relatively vague and ambiguous term, as different practitioners delineate its role, scope, and long-term value in a variety of different ways. However, they all have in common a presumption that there is a problem in the successful transition of new technologies and therapies from the laboratory bench to the patient at the bedside, and that a range of more translational activities (many publicly funded) will be critical to solving the innovation challenge. This has led to significant public resources being channeled into downstream drug development processes, actively reshaping the health innovation ecosystem and its constituent value chains. Although the basis of some of these concerns about the innovation pathway, and the benefit and value of TM as an organizational strategy, is open to some debate, TM's key feature and enduring legacy might in the end be its long-term effect on institutional and organizational practices and the constitution of new types of value, and valuation practices, within the innovation ecosystem.

In the following chapter, I explore in more detail how this amorphous concept is actively reshaping health innovation systems and the conventional everyday practices therein, particularly in the context of interdisciplinary and cross-sectoral collaborations between the commercial and public sectors, which I describe as novel experiments in organizational restructuring. Many of the broader issues tentatively raised in this chapter will now be further unpacked and analyzed.

Chapter 4

Organizational Transformations and the Value of Interdisciplinarity

Introduction

The emergence of new biology as the twenty-first century "big science" (Vermeulen et al., 2013), and the concomitant policy and industry responses to the supposed broken middle of research and development (R&D), has transformed the health innovation ecosystem and the values and R&D practices within it. The implications of the arguments I set out in the previous chapter is that Translational Medicine (TM) is both a general philosophy for "doing applied life science" and a set of specific scientific and clinical activities orchestrated within new institutional settings and interdisciplinary configurations. Although TM is a rather vague and messy concept, as many practitioners are quick to concede, it has had a material effect on the range and type of options and strategies available for therapy development. This has in turn shifted the nature and locus of value within the broader health bioeconomy.

Evidence for the increasing value ascribed to translational approaches can be seen in the growth of collaborations, particularly between academia and the pharmaceutical industry through public-private partnerships (PPPs) or product development partnerships (PDPs), building on the concept of the triple helix of university-industry-government relations (Etzkowitz, 2008). These kinds of organizational experiments, some of which are highly innovative, have become prominent over the past 20 years as the previously indomitable pharmaceutical industry has been compelled to explore alternative routes for innovation and knowledge capture. This has in turn created a much broader role for various public sector institutions and organizations in downstream drug development processes. These include funding agencies, charities, health services, and not-for-profit enterprises. Not all have enthusiastically embraced this trend, because of the high financial risks associated with drug development. Also, some may not consider it acceptable for public money to be invested in projects with such a low rate of success, and question where the ultimate public value is in these types of investment. Others, however, have argued that greater involvement by the state in innovation processes may be needed in contexts where markets have simply failed (Mazzucato, 2013).

In the previous chapter, I unpacked the contested definitions and framings of TM and revealed some of the TM-inspired organizational and policy/funding initiatives that have begun to reshape the therapeutic R&D landscape and engendered new kinds of value. I also introduced the concept of biomarkers as central to the broad TM strategy. In this chapter, I dig a little deeper and interrogate the impact of the translational policy agenda and the demands of the health bioeconomy on R&D practices and knowledge dynamics within the laboratory and the clinic, and between public and commercial organizations. The key question is: *In what ways has the "doing" of R&D been reshaped by the institutional and organizational restructuring precipitated by translational policies and how are stakeholder expectations and values recognized and managed?*

If interdisciplinary knowledge and research, which has evolved within these collaborative partnerships and created new institutional ecologies, disrupts traditional professional and sectoral boundaries, it also reveals uncertainty and tension about where value and benefit might be realized at various locations within the innovation ecosystem. TM and other key factors driving the health bioeconomy envisage new opportunities for therapy development, but they must also confront institutional constraints, organizational conflicts, and the need to manage competing expectations and promissory visions about the proclaimed benefits. My objective in this chapter is to critically reflect on the impact the translational turn is having on therapeutic R&D and interdisciplinary knowledge and practices at the very interface of the laboratory and the clinic. I do this by way of case examples of specific PPPs—the Biomarkers Consortium in the United States, the Translational Medicine Research Collaboration (TMRC) in the United Kingdom, and the Center for Translational Molecular Medicine (CTMM) in the Netherlands—and drawing on interview data with key stakeholders. I argue that although there are challenges facing all kinds of interdisciplinary and collaborative research, and there is palpable skepticism about the novelty of recent organizational and institutional change in the life sciences, we are witnessing the entrenchment of a diverse set of new practices and notions of value within contemporary health R&D.

Interdisciplinary Practices in the Laboratory and the Clinic

The conventional, but somewhat unrealistic, view of technological innovation is that it is both linear, as discussed in chapter 3, and discipline centered. For the physical sciences that dominated the twentieth century, this characterization more or less captured practitioner's accounts of their everyday R&D practices. The general use of technology readiness levels (TRLs), which were created by the National Aeronautics and Space Administration (NASA) and further developed by the Defence Advanced Research Projects Agency (DARPA) in the 1970s, implied compartmentalized and linear stages of R&D. It also suggested that the physical sciences could be understood, organized, and managed according to crude, sequential metrics.[1]

Therapeutic innovation has also been presented as if it follows a linear pathway with R&D stages organized into distinct, discipline-based specialities with their own internal intellectual integrity and broadly recognizable boundaries. However, advances in new biology and the resurgence of concern about translational gaps between the laboratory and the clinic have emboldened many practitioners and adherents of "interdisciplinarity." Indeed, in the 1940s, the American geneticist and later Nobel prize winter G. W. Beadle recognized the problem of disciplinary boundaries in life sciences when he stated:

> It is a most unfortunate consequence of human limitations and the inflexible organizations of our institutions of higher learning that investigators tend to be forced into laboratories with such labels as "biochemistry" or "genetics." The gene does not recognize the distinction—we should at least minimize it. (G. W. Beadle cited in Klein, 2000)

I include this quotation in recognition that the call for interdisciplinarity is not a uniquely modern phenomenon. Nevertheless, crude institutional and disciplinary divisions (if they ever truly existed) have given way to a much greater exaltation of interdisciplinarity as an instrument for sharing complimentary knowledges, institutional and organizational resources, technologies, and expertise to solve specific problems in health-care innovation (McCarthy, 2004). The interdisciplinary turn in the health-related sciences, which has gained momentum since the mapping of the human genome, throws into sharp relief the many barriers that are presumed to inhibit, rather than advance, the translation of basic science into valuable therapeutic products.

Defining Interdisciplinary Research and Its Importance for the Health Bioeconomy

So what is interdisciplinarity? There are various definitions and types of interdisciplinarity, but its emergence as an important policy instrument and way of theorizing and organizing research activities can usefully be linked to Gibbons et al.'s (1994) typology of Mode 1 and Mode 2 research. For Gibbons et al., Mode 1 research broadly corresponds to the hegemony of traditional disciplinary specialisms with their own internal sense of hierarchy, or "orders of worth" to borrow Stark's (2009) terminology. It is exemplified by the notion of an autonomous scientist or institution driving knowledge production with a tightly defined scope and within relatively impermeable intellectual boundaries. Mode 2 research refers to what Gibbons et al. call the "new production of knowledge," which cuts across conventional disciplinary boundaries in order to try and solve complex societal or scientific issues. It is exemplified by diverse and fluid personal and organizational networks of collaboration and is not so narrowly delimited in terms of scale or scope, in contrast to Mode 1 research. So Mode 2 interdisciplinary research is very much context driven and problem oriented. It is also generated in the context of its specific application, usually a particular research problem, as Rekers and Hansen (2015) capture when they talk about complex social problems, such as climate change and aging societies, as grand challenges. These grand challenges require interdisciplinary consortia to both generate the requisite knowledge and challenge conventional, "business as usual" approaches (Rekers and Hansen, 2015: 244). The problem of translation in the health bioeconomy could also be described as a grand challenge that requires these interdisciplinary approaches to R&D, and recognition of the multiple and complex pathways to the clinic, each of which brings its own particular opportunities and hurdles. Indeed, the very idea of interdisciplinarity is often inextricably linked to the notion of complexity, which Klein (2004) has described as an "evolving relationship."

Gibbons et al. argue that Mode 2 research began to take root in the second half of the twentieth century. Although, as we saw in previous chapters, there were elements of interdisciplinary (or transdisciplinary) practices and collaborations much earlier in the context of the biological sciences. This was exemplified in the work of Pasteur in the nineteenth century and Banting and Best in the early twentieth century. Nevertheless, there is a compelling argument that the late twentieth century witnessed a much greater disruption to the conventional process of discipline-oriented research and the autonomy of the actors and organizations conducting it. In her account of the emergence of biomedicalization, Lowy (2011) argues that although doctors had slowly come to embrace laboratory sciences in the nineteenth century, the use of the term "biomedicine" as shorthand for the work of both doctors and scientists, and the homogenization of methods for both fundamental life sciences and applied clinical investigation, came to predominate in the 1920s and 1930s. As I briefly mentioned in chapter 3, World War II was seen as a turning point in this biomedicalization process and accelerated collaborative research in the life sciences (Lowy, 2011: 117).

An excellent example of this kind of Mode 2 interdisciplinary research in the biological sciences is described by Lyall et al. (2011). They point to the discovery of the structure of deoxyribonucleic acid (DNA) by Watson, Crick, Franklin, and Wilkins as exemplifying the contrast between interdisciplinary working practices that coalesce around a core research problem, with the conventional and more mundane disciplinary approach. The authors argue that Franklin and Wilkins were discipline-oriented researchers focused very much on achieving a specific goal and tried to avoid getting distracted by alternative approaches or modes of enquiry. In contrast, Watson and Crick were "intellectual butterflies" who sought relevant knowledge, information, and data wherever they could find it. By taking two key insights from the chemists Erwin Chargaff and Jerry Donohue, and combining it with data from the work of Franklin and Wilkin, they were able to determine the double helix structure of DNA (Lyall et al., 2011: 7).

So interdisciplinary research can be defined as the integration or synthesis of perspectives at the very interstices of disciplinary knowledge and practice. This is how Calvert (2010) defines it, in contrast to "multidisciplinary research," where several disciplines are brought together to solve a particular research problem, but each retains its core disciplinary integrity and there is very little cross-fertilization between the contributing disciplines (Tait and Lyall, 2007). It can also be contrasted with "transdisciplinary research," where specific issues or theories from established disciplines, rather than the disciplines themselves, are used to solve new and emerging research problems. Interdisciplinary research can also be "academically oriented" or "problem focused," with the latter defining most of the interdisciplinary initiatives created in response to the multiple challenges facing health translation.[2]

Calvert (2010) also distinguishes *individual* and *collaborative* interdisciplinarity, arguing that this important distinction is rarely discussed in the literature. Individual interdisciplinarity refers to cases where an individual researcher "integrates perspectives from different disciplines in their work," and collaborative interdisciplinarity is the result of a "collaborative endeavour, where different disciplines come together to bring their insights to a problem" (Calvert, 2010: 202).

So why is interdisciplinary research considered so important in meeting the challenges of health innovation and supporting the bioeconomy? I suggest that the complexity problem in the biological sciences, and the diverse range of expertise, skill sets, technology, and knowledge domains that are now central to biomedical innovation, necessitates the cultivation of interdisciplinary research teams and organizational structures. Indeed, by presenting the interface between the laboratory and the clinic as the key site for policy intervention, the translational turn in therapeutic innovation places interdisciplinarity at the very center of its remit.

Funding agencies are demanding that there be much closer collaboration within and between the natural and social sciences to advance science, technology, and innovation for broader societal benefit. Some argue that funding agencies are crucial in creating interdisciplinary knowledge, and without their support and tacit knowledge of managing interdisciplinary research programs, the approach could lose its transformative potential (Lyall et al., 2013). If complex problems cannot simply be resolved within the structures of conventional disciplines working in isolation, support for more open and fluid boundaries is required. This may necessitate the development of new methods, evaluation cultures, and organizational restructuring, which is highly disruptive to institutions that have been built up over centuries to compartmentalize knowledge and manage career development according to relatively circumscribed disciplines. Furthermore, Robertson et al. (2003) suggest that while the benefits of interdisciplinarity are well acknowledged, the methods of "...interdisciplinary collaboration are opaque to outsiders and generally remain undescribed" (Robertson et al., 2003: 2). I explore this aspect in more detail later.

Institutional Support for Interdisciplinarity

In recognition of the increasing value of interdisciplinarity in driving health innovation and the bioeconomy, a number of public sector institutions and commercial research organizations have implemented interdisciplinary programs and resourced infrastructure as part of a broader translational agenda. In the United States, some of the National Institutes of Health (NIH) Road Map Initiatives, which I described in some detail in the previous chapter in the context of bridging the laboratory and the clinic, include a major focus on capacity building for new interdisciplinary research contexts.

The NIH Roadmap Initiative explicitly set out to encourage scientists to use all available technologies and databases to go beyond their individual disciplines and explore complex biological systems as members of new interdisciplinary scientific teams. Crucially, partnering was encouraged not only with other academic centers, but also with the commercial sector. The ethos of the Roadmap, and the NIH Common fund, was to build interdisciplinary research teams of the future. A key innovation was the expectation that scientists would collaborate with physicians in one dedicated building. This was an explicit acknowledgment that interdisciplinary research requires new infrastructure and architectural design to facilitate collaboration and knowledge exchange. Interdisciplinary translational research is far more than just science, technology, and people. It is also about bricks and mortar. In the context of its general Interdisciplinary Research program (IR), the NIH sought to change academic research culture to benefit from the dynamism of interdisciplinary research. The IR program included initiatives to dissolve conventional departmental boundaries within academic institutions and routinize interdisciplinary ways of working.

This particular initiative led to the creation of nine Interdisciplinary Research Consortia to create more integrated projects in terms of basic and applied science, training, and the organization and management of administrative structures. This may be seen as an explicit attempt to normalize interdisciplinarity and create the space and institutional support for it to flourish. The institutional dimension of interdisciplinary research cannot be overestimated. As Robertson et al. (2003) argue, institutions are important in the: "...building of an institutional 'platform' for collaboration: an infrastructure of research organizations, academic journals, funding committees and informal networks of researchers that actively foster interdisciplinary research" (Robertson et al., 2003: 2). The authors proceed to argue that in the1950s, the Rockefeller Institute of Medical Research (now the Rockefeller University) was one of the first to take seriously the need to bring together a range of different sciences and dissolve traditional disciplinary boundaries, both intellectual and cultural. This led to the Institute's significant biomedical breakthroughs throughout the 1950s, 1960s, and 1970s (gene regulation and the structure of antibody molecules being two notable examples). However, the interdisciplinary approach was considered radical for the time and even today there are persistent barriers to interdisciplinary research and practice. Although there are now undoubtedly more funding opportunities for interdisciplinary research in the biosciences, and there has been real organizational change precipitated by its underlying philosophy, the key question is how are everyday R&D practices being transformed and valued by different stakeholders and practitioners?

Interdisciplinarity and How It Shapes R&D Practices

One field that nicely illustrates the expectant value of interdisciplinarity, and the need for translational R&D practices, is clinical pharmacology. According to many of its practitioners, clinical pharmacology provides the underlying knowledge, tools, and skills required for the full realization of TM. In 2007, for example, the United Kingdom's Wellcome Trust (a charitable funder of biomedical science) stated:

> There is an urgent need to develop individuals who have the ability to combine a firm grounding in the principles of basic and clinical pharmacology with the most modern research technologies to address complex (patho) physiological questions. Such individuals will play a key role in shaping the interdisciplinary research that underpins translational medicine and therapeutics. (cited in Aronson et al., 2008: 154)

Aronson et al. (2008) argue that clinical pharmacology is an exemplar science for the two-way process of bench-to-bedside research, as it encompasses everything from molecular drug discovery to research in the use of therapies in individuals and populations (Aronson et al., 2008: 154). The authors consider the interdisciplinary ethos of clinical pharmacology to be a major factor in realizing the long-term value of, for example, biomarkers and the associated development of personalized/stratified medicines.

Many of my interview respondents also expressed strong views about the value of interdisciplinarity in the context of translational health innovation, particularly those who worked in policy and regulatory settings. One respondent from the policy side gave the following illustrative example:

> When you undertake sophisticated MRI or PET scanning as part of the translational activity, not only do you need the scanners themselves, you also need the radiographers and the technicians and the statisticians who will analyze the images, but you also need the academic translational leaders to actually develop the technology and the

use of imaging in particular areas...not only is it a team sport but it's actually getting skills from a whole breadth of different areas to address the key issues. (Director of Policy 2, Funding Organization)

This account implicitly focuses on the need to enhance the laboratory/clinic interface by facilitating the integration of complementary skills and tacit knowledge. However, interdisciplinary research does not always occur naturally by simply bringing together several disciplines in a research project. Institutional support or infrastructure is also needed to facilitate the formation of a cohesive research team involving researchers from different disciplines, combine expertise from several knowledge domains, and overcome communication problems among researchers from different disciplines (Lyall et al., 2011).

Another illustrative example is provided by systems biology and pathway medicine, where it has proved difficult to establish research within conventional academic structures. The aim of systems biology is to generate and integrate "big data" sets using computational and mathematical tools. The ultimate goal is to accurately model biological systems in silico. To accomplish this requires the complimentary skills of physicists, computer scientists, engineers, mathematicians, and biologists. Calvert (2010) describes systems biology as interdisciplinary in the collaborative sense. The motivation for interdisciplinarity in this context is that the data sources are so complex and diverse that true understanding, and application, requires the successful integration of different disciplines, knowledge domains, technical skills, and expertise. However, traditional academic environments can make interdisciplinary research difficult to operationalize. Some leading systems biologists have challenged the orthodoxy and the constraints of academic bureaucracy, and embraced the notion that interdisciplinarity requires novel building design and new types of infrastructure to facilitate the very different R&D practices that are required. Calvert (2010) has observed that systems biologists often work in buildings with no walls between the laboratories and social spaces that encourage the "wet" and "dry" scientists to communicate.

However, even with infrastructure to facilitate interdisciplinarity, cultural differences between disciplines can be a perennial source of friction. Kling (2006) has observed this in relation to biologists and computer scientists:

Biologists think of themselves as wise, sagely knowledge banks, and they see computer people as keyboard jockeys. The computer guys think of themselves as mathematics-driven scientists. They think of biologists as lab technicians. (Kling, 2006: 1306)

Kling argues that these attitudes must change if systems biology is to succeed and deliver real clinical value. Different working methods, underlying assumptions about the nature and interpretation of data, and cultural attitudes toward the core research problem can make interdisciplinary working relationships difficult. According to Kling, ideas about what it is to do good science, and what science should ultimately aim to achieve, are regularly discussed and debated among the participants in interdisciplinary teams.

This idea of breaking down physical barriers, and designing the organizational infrastructure and operationalizing institutional norms in such a way that they facilitate interdisciplinary research and serendipitous knowledge exchange, was also expressed by one inter-view respondent working in the United Kingdom's National Health Service (NHS). She stated:

> I think another aspect that's important is finding new multidisci-plinary links, and those come out of casual conversations as much as from formal structures, so they're a little harder to control and direct. But once they happen, providing support for them is most important. (Senior Health Services Manager 3)

Here, emphasis is on the role of informal interaction as a way of pro-moting interdisciplinarity, although the respondent recognizes the importance of institutional infrastructure to guide and capitalize on inchoate cross-disciplinary ideas. This notion of internal knowledge transfer has long been recognized and exploited by the pharmaceuti-cal industry, as we saw in chapter 2 when discussing serendipitous "knowledge spillovers" within multinational firms (Henderson, 2000). The drive toward promoting interdisciplinarity and the break-ing down of conventional professional boundaries continues apace among those promoting TM as the answer to the challenges of health R&D and the needs of the bioeconomy. The question then is what real impact does this have on actual R&D practices, and what chal-lenges arise from attempts to institutionalize collaborations and ini-tiatives across sectors, disciplines, and professions?

TM and Shifting Professional Boundaries in the New Health Bioeconomy

The TM philosophy of "bench-to-bedside and back again" requires a disruption of conventional, rigid boundaries between disciplines and laboratory and clinical spaces. It also challenges existing institutional and professional norms, and may require fundamental change in the relationship between public and commercial spheres. The feedback loop from clinic to laboratory has perhaps been the most difficult to identify in biomedical practice, because its role and significance has often been marginalized by a focus on the bench-to-bedside continuum. Here, the emphasis has been on the need to push innovative science into the clinic, with policy initiatives targeting the front end of the research pathway. Some authors suggest that this is partly due to a "hierarchy of credibility," where clinicians have historically enjoyed less privileged status than academic scientists. The scientists have therefore tended to define the key problems of translation and set the parameters of candidate solutions under the auspices of TM. According to Hoonaard (2009), clinicians might actually be complicit in sustaining this hierarchy by privileging the discourses and practices of scientists.

Resurgent interest and debate about the role of the so-called "clinician-scientist"[3] is interesting in the context of this problem concerning shifting professional boundaries and perceived tension or conflict at the laboratory/clinic interface. Are new alliances, relationships, and practices at this interface crucial to the success of TM, especially in the context of emergent biotechnologies that rely on interdisciplinary expertise? Wilson-Kovacs and Hauskeller's (2012) account of the clinician-scientist in stem cell research reveals that professional legitimation may be achieved and consolidated by professionals able to better link the laboratory and the clinic. They suggest that conventional professional hierarchies are being actively reshaped by TM policies and the challenges of life science research. Martin et al. (2008) also draw on the concept of "communities of promise" to facilitate understanding of what they describe as new configurations between clinical and basic research that have coalesced around particular sociotechnical objects in stem cell research. These authors point to the ways in which the "stem cell story" has centered on expectations about how clinical developments emerge and idealized assumptions about the distinctive roles and value of basic science, clinical science, and the commercial sector. The discourse of translation and interdisciplinarity, according to these authors, has thrown into sharp relief the "more complex and dynamic relationship between the spaces and communities of science and application in the clinic" (Martin et al., 2008: 30).

My interview accounts substantiated many of these views by revealing lingering tensions between the different professions involved in

TM and uncertainties about the new institutional dynamics demanded of contemporary life science R&D. On the issue of the clinician-scientist, for example, a senior academic respondent stated:

> There was really a big emphasis on developing clinician-scientists a number of years ago but I think the pressures of medicine and bureaucracy and the demands of running a laboratory, the increasing expectations on what you can do in science to be competitive and to get research funds has made this role unattractive. But this role is critical to the discipline of translational medicine. Other expertise is important as well but I see this group in particular being at risk. (Senior Academic Scientist 5)

This respondent presents the broader institutional environment and vicissitudes of competitive science as a threat to the new professional alignments demanded by TM. How can the laboratory and the clinic be better bridged, and interdisciplinary research supported, when there are persistent cultural, institutional, and intellectual challenges manifested at this interface? Many respondents argued that basic scientists are often naïve about drug development processes and clinicians are often too distant from the science to be able to contribute in any meaningful way their clinical knowledge and expertise. A senior clinical academic stated:

> I would say if you looked at the basic scientists there is a naivety about how to develop drugs and how to apply them clinically, in some cases complete ignorance, and even people who are considered to be experts I don't consider to be so. At the clinical end, I think there is a difficulty in terms of the commitments that clinicians have in an academic environment, obviously teaching, research and patient care is a major impediment [to contributing to TM]. I would certainly feel there is definitely room for expanding the academic investment in training and in creating positions in translational medicine [such as clinician-scientists] and making it a discipline in its own right. (Senior Clinical Academic 7)

Advances in new biology and the growing demands of the bioeconomy have led to a much greater focus on capitalizing on new R&D practices at the laboratory/clinic interface and disrupting conventional professional and disciplinary boundaries. Although, as I stated in the previous chapter, many of the R&D challenges and translational approaches are not new, the emergence of new biology as the exemplary big science of the twenty-first century has shaped what it means to do contemporary health R&D. This becomes much clearer in the context of public/private collaborations, which must organize interdisciplinary science across institutions and sectors and manage very different expectations of value and benefit. I now discuss public/private collaborative research projects as a new organizational regime for the health-related life sciences which, despite challenges common to all types of collaborative work, are essential to deliver the broad benefits from new biology.

Public/Private Collaboration as a New Organizational Regime

So far I have considered the challenges of bridging the laboratory and the clinic and institutionalizing interdisciplinarity, which disrupts professional boundaries as well as research and clinical cultures. However, TM also involves more substantive cross-sector collaborations, particularly between academia and the pharmaceutical industry. This has been the most resource-intensive application of TM and has involved the establishment of PPPs and PDPs and the increasing role of public sector finance and expertise in downstream drug development. There are many examples of such initiatives. In the United States, as well as the heavily resourced National Center for Advancing Translational Sciences (NCATS) discussed in the previous chapter, the NIH Foundation manages the Biomarkers Consortium, which is a large public-private research partnership aimed at identifying and validating biomarkers for drug development, preventative medicine, and diagnostics. In the United Kingdom, there has been the TMRC in Scotland (with investment from Wyeth pharmaceuticals and Scotland's development agency Scottish Enterprise), while in the Netherlands there has been the CTMM, which involves multiple public sector and commercial research and clinical organizations collaborating to develop technologies and tools for personalized medicine.[4]

It is within these new collaborative structures that the values and expectations of different stakeholders are being made to converge and redefining what it means to "do R&D" in the twenty-first century bioeconomy. In this section, I briefly describe these three public-private collaborations as exemplars of a new organizational structure for R&D, before examining stakeholder views about collaborative R&D and the opportunities and challenges it presents. This reveals the importance of adopting a broader and more nuanced approach to value, valuation practices, and the management of expectations. However, before I describe these three exemplar PPPs, which have emerged in wealthy, industrialized nations as means to capitalize on the opportunists and challenges of the health bioeconomy, I want to briefly highlight the important role PPP's have played in developing counties, particularly in Africa.

PPPs and the Broader Notion of Value in Precompetitive Research Collaborations

The necessity and broader value of PPPs were perhaps first recognized in the context of low-resource countries and regions where market failure has been blamed for the lack of effective medicines, especially vaccines, for many tropical diseases (Hanlin et al., 2007; Mugwagwa et al., 2013). Lezaun and Montgomery explore the rise of the PPP and associated emphasis on "open innovation" as having particular resonance in the context of neglected tropical diseases. The authors suggest that a new moral economy of R&D has emerged through these PPPs, or what are sometimes referred to more narrowly as PDPs, between pharmaceutical firms, nongovernmental organizations (NGOs), and charities, such as the Bill and Melinda Gates Foundation.

The authors argue that these kinds of partnerships have created new circuits of exchange as intellectual property (IP) is pooled, along with compounds, data, and expertise, such that it shifts from being

> an instrument of exclusion...to a mechanism of attraction, a bait used to draw potential partners into collaborative research agendas. IP still functions as an instrument of control, but its mode of operation changes. (Lezaun and Montgomery, 2014: 4)

In this context, the valency of property rights, which historically have been used to determine economic value and demarcate public and private spheres and interests, dissolve conventional boundaries and allow different stakeholders and actors to coalesce and realize broader notions of value and benefit. This kind of open, precompetitive innovation model is considered essential for the development of innovative therapies in complex technology areas and sociopolitical contexts characterized by risk and uncertainty. Here, risk and uncertainty refers to both the basic science and technology (i.e., does it work?) and the various pathways to clinic (i.e., is there a route to market and the appropriate health-care infrastructure to deliver the product?). In these contexts, research silos are incapable of delivering viable therapeutic products to market. Consortia built on principles of open innovation are therefore preferred.

PPPs are considered a model of good practice for delivering new therapies in low research settings and are interesting because they embrace broader notions of value (beyond the economic) in response to conventional market failure. The question is what are the benefits and challenges of applying this model to a developed country context? Are stakeholder expectations of different but complimentary benefits accruing from participation in collaborative partnerships being sufficiently realized and appropriately managed in industrialized nations? What are the key challenges to making these new organizational regimes work in practice? Rai (2005) argues that while it is not yet clear that open and collaborative models of innovation will engender more socially desirable and valuable innovation in the biomedical sciences, either in absolute terms or relative to models based on exclusionary IP regimes, the approach is still probably worth pursuing. In the context of bioinformatics, he argues there is excitement that "wet lab" systems biology, developed under an open innovation regime, might provide a more coordinated response to the complexity problem than "small-lab" biology (Rai, 2005: 133).

With this in mind, I now describe three exemplar case studies of PPPs, each focused on biomarkers but with slightly different organizational structures, modes of operation, and scope of value and benefit.

The Biomarkers Consortium

The Biomarkers Consortium was launched in 2006 as a PPP centered on identifying and validating novel biological markers in four key disease areas: cancer; inflammation and immunity; metabolic disorders; and neuroscience. It is a model of what is commonly referred to as "precompetitive collaboration," which operates with the underlying philosophy of "open innovation," including the use of "open source" IP.[5] In terms of IP, all partners involved in a specific project funded by the consortium must grant other partners in the project access to any preexisting data and resources relevant to the project on a limited, nonexclusive, and royalty free basis. This is for research purposes only, and project participants cannot gain proprietary rights to another member's preexisting IP simply from participating in the project. In terms of generating new data and IP, the nonfederal participants (meaning anybody not employed by a federal agency) do have a right to protect inventions through IP, but they must provide a nonexclusive and remuneration-free license to the other project participants and a nonexclusive research license to all other members of the consortium. Wagner et al. (2010) provide a nice and concise explanation of this kind of open source philosophy when they write:

> Precompetitive collaboration, defined as competitors sharing early stages of research that benefit all, is a potential driver for innovation and increased productivity. Open-source software research has grown from a cottage industry into a business model, with products such as Linux becoming mainstream alternatives to industry leaders. Biology has evolved into an information-rich science, and the analogy with software could provide a potential answer to productivity issues for biomedical research and drug development. (Wagner et al., 2010: 539)

The authors consider initiatives such as the Biomarkers Consortium, which build on rich networks of public and commercial organizations operating under the principles of "open source" that are well established in the IT sector, as a significant and increasing trend in biomedical R&D. The Innovative Medicines Initiative (IMI) in Europe is a similar kind of program that aims to support innovation in precompetitive environments, with the expectation that there will be major downstream benefits for both public health and the broader bioeconomy.

The Biomarkers Consortium is managed by the Foundation for the National Institutes of Health (FNIH), and also involves the NIH and FDA, the Centers for Medicare and Medicaid Services, the Biotechnology Industry Association and the Pharmaceutical Research and Manufacturers of America, as well as public and patient advocacy organizations (Zerhouni et al., 2007: 250). The commercial partners provide most of the financial support for the Biomarkers Consortium, while the public-sector partners provide in-kind contributions. The Biomarkers Consortium is also funded through membership fees, paid by the commercial partners,there is a clear translational element. It is not simply about the identification and validation of biomarkers. The involvement of regulatory bodies, health-care payers, and public sector NGOs highlights the fact that clinical utility and social value are considered an integral part of the mission. Altar (2008) argues that the fact that the information is quickly deposited in the public domain reinforces the precompetitive nature of these initiatives (Altar, 2008: 361). The argument is that by sharing costs, resources, expertise, and risks associated with this kind of research, the challenge of validating clinically relevant biomarkers and getting them into routine practice can eventually be overcome. As I described in the previous chapter, novel molecular biomarkers are a highly promising but also complex area for biomedical innovation. The value of the Biomarkers Consortium lies in its ability to rapidly turn over projects and ensure results are made publicly available, and there are perceived to be multiple benefits to all partners involved.

Industry benefits from having qualified biomarkers for use in drug development programs, and regulators benefit from the development of new tools through which to assess therapeutic safety and efficacy and therefore improve regulatory science. However, Zerhouni et al. (2007) believe that the ultimate benefits will be for patients and the general public. As in the previous chapter, we can see here again the building of future, promissory visions of public value, and clinical benefit. There is an expectation within these kinds of consortia that the cumulative value of many small-scale, precompetitive projects will, in the future, bring major benefits to science, patients, and the bioeconomy. Of course, the benefits are often broad and vague and couched within the rhetoric of "better" and more "efficient" drug development. In the broader context of big science projects in biology, Davies et al. (2013) have questioned the rhetoric and hype that tend to drive many of these kinds of projects:

> Rhetorics and practices of data sharing, standardisation and milestone setting mobilise and aggregate biological properties and capacities at different scales, through different means and with different effects. Some of these projects may be bigger and faster than what came before, but the question of whether they are necessarily better is more openly contested. (Davies et al., 2013: 391)

The question is whether PPPs such as the Biomarkers Consortium are "anticipatory" or "promissory" organizations that build lofty and idealistic expectations, and envisage technological futures that simply reinforce the untested belief that this is a model of best practice. Alternatively, are these kinds of collaborations a necessary response to the challenges of contemporary drug development which, on their own, may have limited or marginal impacts but when aggregated could have a more transformative impact on the health bioeconomy?

In terms of what the Biomarkers Consortium has delivered, Wagner et al. (2010) describe the first successful project, which was in the field of metabolic disorders. The project sought to evaluate the "utility of adiponectin as a marker for glycemic efficacy through pooling existing data from clinical trials from multiple sponsors" (Wagner et al., 2010: 54). The project was initiated in 2007 and completed two years later. The project analyzed blinded data on more than 2000 diabetes patients from randomized and placebo-controlled trials from four pharmaceutical firms: GlaxoSmithKline (GSK), Lilly, Merck, and Roche. The companies Quintiles and NIDDK were responsible for the data analysis. The project was successful in confirming the suspected link, and demonstrating the utility of adiponectin as a valid marker/predictor for glucose tolerance in both type 2 diabetic patients and healthy individuals. According to the authors, despite the many challenges facing this kind of project, it was successful in generating answers to important questions that would have been impossible to resolve using only the data sets of the individual companies. As of 2014, 11 projects have been completed within the Consortium and 8 are ongoing.[6]

The Translational Medicine Research Collaboration[7]

The TMRC was established in 2006 with the ambitious aim to create an internationally recognized scientific and clinical research network in Scotland focused on identifying and validating biomarkers and diagnostic testing for new drug development. At the time, it described itself and was generally regarded as the first large-scale collaboration between industry, government, and academia, with an initial budget of approximately 80 million USD for a five-year research program. The collaboration involved Scotland's major research universities and medical schools (Aberdeen, Dundee, Edinburgh, and Glasgow), the cities' regional health boards, Scotland's economic development agency, Scottish Enterprise, and the multinational pharmaceutical firm Wyeth, which was headquartered in the United States (Mittra, 2013b). Wyeth was eventually purchased by Pfizer in 2009, and the collaboration did not continue into a second round of funding, as I describe below.

The impetus for setting up the TMRC was Scottish Enterprise's desire and vision for a major collaboration in Scotland based on its life science assets and expertise. This, it hoped, would generate revenue, jobs, and further inward investment from the pharmaceutical industry. Scottish Enterprise's vision fortuitously coincided with Wyeth's interest in setting up a PPP focused on TM, which it had initially sought to do in the United States, but failed to identify optimal opportunities that would provide access to sufficient patient data and tissue samples. The basic idea was to set up a translational initiative very much focused on biomarker development to tackle the challenges facing the middle stages of the R&D process, particularly phase 2 attrition rates for drug candidates. So TMRC evolved from this shared desire and commitment of a major company and a public regional development agency. Both became co-funders of TMRC. Each committed core funding for the first five years of the Initiative (Wyeth committed 53.5 million USD and Scottish Enterprise 33 million USD). The primary role of Scottish Enterprise was to be an investor and facilitator for the collaboration, with Wyeth being a more active partner in driving some of the projects. Wyeth wanted to use the collaboration to gain access to basic science and clinical resources to discover and validate novel biomarkers so that it could better predict the chance of success for many of its phase 2 drug compounds.

TMRC involved the establishment of a central research laboratory at the University of Dundee, which housed a range of technologies, tools, and expertise relevant to biomarker studies. This laboratory was linked to the other academic centers and the health services. The important link to patients and clinical samples, through the NHS, was a major driver for the collaboration. The NHS in Scotland is highly integrated, with fully electronic patient records and stream-lined ethical approval systems for patient studies, so Wyeth saw this as a valuable resource. The metaphor of "bench-to-bedside and back again" captures the relationship that was anticipated between the academic centers and the health service partners.

Overall coordination and strategic management of the PPP was conducted by TMRI Ltd., which was the company established to act as the delivery mechanism for the collaboration. It was responsible for distributing funding to the individual projects and evaluating and marketing the IP generated from the studies. Like the Biomarkers Consortium, there were shared IP agreements in place, although unlike the Biomarkers Consortium, the commercial value of the IP would ultimately reside with Scottish Enterprise and Wyeth. This was not a conventional, large-scale precompetitive initiative. There was a clear commercial drive and imperative that had to be satisfied. Wyeth, for example, would retain IP ownership on work directly related to its drug compounds, with other drug-related IP being shared equally with Scottish Enterprise. The latter would, however, retain IP on any diagnostics and tools developed during the collaboration. This is where the public sector investor in TMRC saw the ultimate value in the initiative.

The expectation, according to people I interviewed who were involved in the TMRC, was that the collaboration would advance developments in diagnostics and therapeutics by combining resources and expertise to improve various aspects of the innovation life cycle. The first round of project calls was initiated in May 2006, when more than 80 proposals were submitted. As many as 65 programs were supported in the first two years. However, TMRC only had guaranteed funding for five years (2006–2011), although there was an expectation that a further five years of funding would be forthcoming. However, in 2011 it was announced that there would be no follow-on funding. Following Pfizer's purchase of Wyeth in October 2009, the future of the TMRC was always going to be uncertain. The TMRC program slowly wound down during 2011 and 2012 as core assets were divested. Despite the fact that follow-on funding was never guaranteed, and there was some promising scientific outputs from the initial investment, the ending of the initiative was a blow to erstwhile Scottish aspirations to be an international leader in TM activities and a hub for translational collaboration in pharmaceuticals. Indeed, the public sector funders marketed the initiative through promissory discourses about national competitiveness and built up early expectations. So TMRC serves as a cautionary tale of how a highly promising PPP, which is one of many organizational innovations considered essential to the realization of value from twenty-first century life sciences, can be usurped by sudden changes in commercial strategy. One of the risks involved in this kind of collaboration was the fact that it was very limited in scope and its success was tied to the fortunes of a single company, which raises questions about how value is shared and expectations of very different institutional partners realized. I discuss this in more detail in the final section of this chapter

The Center for Translational Molecular Medicine

The third of my case examples is the CTMM in the Netherlands. CTMM is a PPP that was established in 2006 (with first round of projects starting in 2008) with the ambitious aim to develop new molecular diagnostics and imaging technologies that could contribute to the realization of personalized medicine. Focusing predominantly on cancer and cardiovascular disease (responsible for two-thirds of all deaths in the Netherlands), and to a lesser extent on neurodegenerative diseases, the CTMM is a broad-based PPP involving multiple companies (over 80) and public sector organizations (universities, academic medical centers, and hospitals). The principal industrial partners at the beginning of the initiative were Phillips, a leading medical technology company, and Organon, the Netherland's largest pharmaceutical company. The chemical company DSM, and FEI, a major producer of electron microscopes, also played a key role in the early development of CTMM (CTMM, 2006). Industry partners were expected to provide the Information Technology (IT) backbone of the collaboration and support the development of bioinformatics and biostatistics necessary for biomarker discovery and validation. A number of smaller, specialized companies would also contribute to the actual research projects funded under the consortium and commercialize any outputs. The academic partners included the Netherlands' leading academic research centers in cancer, cardiovascular disease, and neurodegenerative diseases. They would be responsible for driving the basic science and providing the academic knowledge and expertise to complement the biomarker studies and diagnostic development.

In terms of outputs and relevance, the business plan of the CTMM stated:

> CTMM will focus on the main causes of mortality and diminished quality of life. This is reflected in the distribution of research funds [half dedicated to cancer and cardiovascular disease]...CTMM will contribute to advancing the level of care as well as to enhancing cost-effectiveness of healthcare measures and approaches (i.e. cost containment). It will meet the need for molecular guided introduction of new treatment modalities through a much more efficient evaluation process with biomarkers and molecular imaging in the interest of the public and in the interest of reducing the costs incurred by large and time-consuming clinical studies. (CTMM, 2006: 24)

This focus on multiple benefits (health, cost-containment, and promoting innovation) aligns CTMM with the Dutch valorization model described in chapter 1 (Stemerding and Nahuis, 2014). Here, Dutch innovation policy has sought to embrace a multiplicity of values, beyond crude and instrumental economic criteria, in supporting these large-scale projects. The policy and practice of valorization was an attempt to establish a new social contract between science and industry. In response to growing concern among many scientists that commercial priorities and values were compromising the broader aims of science and its broader public value, valorization emerged as a national strategy to redress the balance and reemphasize the broader range of values embedded in innovation processes. Although the authors claim that valorization initially prioritized economic impacts of science, other values did, in theory if not always in practice, later come to frame innovation policy within the country. The CTMM in many ways reflects the spirit of this policy, as the different but complementary interests and values of health-care systems, payers, and patients, as well as academic scientists, were considered an integral part of its mission.

My interviews with senior representatives from the CTMM revealed both the novelty of the initiative as well as some of the major challenges with these kinds of partnerships. Starting with the initial development of the PPP, one respondent described how three key groups with an interest in translation drove the initial proposal. The first was the commercial company Phillips Healthcare, which had a strong interest in establishing direct links with clinical academic centers. Phillips wanted the academics to help it choose the correct technological applications that would translate into improved patient care (bedside-to-bench translation). This was a similar objective to Wyeth when it established the TMRC. The second group was the oncologists. This group was entirely driven by the notion of improved patient care rather than the technology options themselves. They had questions derived from their experience in patient care and wanted to link with technology providers to accelerate the translation of new diagnostics and therapies towards the clinic (bench-to-bedside translation). A third, but smaller, academic group was interested specifically in Alzheimer's disease and wanted to develop better and earlier diagnosis to facilitate patient care. They wanted to explore new platform diagnostic technologies to drive new and exploratory drug development.

These three groups teamed up and sought support from the Dutch government for the establishment of a public-private consortia. What they successfully attained was direct government funding of 150 million euros over 5 years, which was matched with 75 million euros in-kind contribution from the academic partners and 75 million euros from industry (half in-kind and half cash). This contribution led to three major proposal calls and the funding of 21 projects (40 percent in oncology, 40 percent in cardiovascular, and 20 percent in other areas, including Alzheimer's disease, rheumatoid arthritis, and infectious disease).

The CTMM was novel, in comparison to conventional funding of research in the Netherlands. First, it was unique in the Netherlands in terms of the public-private collaboration and the cross-fertilization of ideas between academia and industry that were enabled by the partnership. Second, the interdisciplinary nature of the projects was considered novel. Over time, the CTMM has attracted many new partners and, as of 2012, involved over 80 different companies (with two-thirds small and medium-sized companies). Like the Biomarkers Consortium, this is very much a precompetitive collaborative arrangement with shared IP arrangements to ensure all partners benefit from participation. On this issue of IP, one interview respondent stated:

IP is crucial...if you make it negotiable for every individual partner then probably this whole thing would never have got off the ground. So what we did was we had an IP working group already before we started and they took a whole year....we said we are going to discuss IP without having an actual programme on the table and work out what kind of IP would work. We had the different parties around the table. We had representatives from two SME companies, one technology company and one big pharma company. We had a lawyer that represented all of the academic centers in the Netherlands. It took them a year to develop a proposal that could be taken forward and was non-negotiable. (Senior Academic Scientist and member of the CTMM)

As in the other two cases described above, IP was crucial to get right at the beginning of the project, with value negotiated and shared such that each partner benefited in some way from being involved. At the very least, each participant was assured that in the absence of short-term economic gains or clinical impact, there would be multiple future benefits.

So these are just three examples of PPPs that have developed in response to the challenges of new biology and health in the twenty-first century, and reflect the new organizational forms that are driving the health bioeconomy. They each reflect in their own ways the challenges of collaboration, but also the diverse expectations and visions of future value and benefit that are co-produced and drive interdisciplinary, collaborative research. The Biomarkers Consortium and the CTMM are both large-scale precompetitive collaborations that were always envisaged to be diverse, long lasting, and able to grow organically. The TMRC was always more narrowly focused and closed in terms of participating institutions and the nature of the research and its use. Nevertheless, all demonstrate how actors, organizations, and institutions can come together with different notional ideas of value and benefit from collaborative research and make R&D work in practice.

The level of collaboration that now takes place in the biosciences is far higher than it has ever been in the past. A recent report from the Tufts Center for the Study of Drug Development states that collaborative research between drug companies and service providers reduces development risks and suggests that more than half of all new drugs that were approved by the FDA between 2000 and 2011 were developed by companies that had collaborated with other organizations (both public and commercial).[8] However, there remain important questions about how stakeholder expectations of value and benefit are managed and realized in these kinds of partnerships. What do people actually involved in these kinds of initiatives believe are the benefits and challenges, and what does this tell us about the nature of R&D in the twenty-first-century health bioeconomy? I now reflect on different stakeholder accounts of collaborative R&D.

Stakeholder Value and the Management of Expectations

As discussed briefly in chapter 1, Pisano (2006) emphasizes the clash of norms and culture between academia and industry as a key challenge for the translation of life sciences into a viable and sustainable business:

> Conflicts between science and business—some obvious, some subtle— are apparent at many levels, beginning with their different cultural norms, values and practices. For example, science holds methodology sacred; business focuses on results. Science values openness and sharing (with attribution); business generally demands secrecy and propriety. Science demands validity (Is this idea/finding valid? Does it stand up to scrutiny?); business demands utility (Is it useful?). Both areas can be fiercely competitive, but they compete for different currency. Science "keeps score" by intellectual impact and contribution to a body of knowledge, as measured by prestige, academic standing, peer evaluation, and published articles; business does so by financial performance. The clash of these norms, values, and practices becomes most apparent when private enterprises and universities collaborate. (Pisano, 2006: 6)

This sentiment was important in all three case examples discussed above, where the management of different expectations, values, norms, and practices was critical to success. Many of my interview respondents had been involved in these kinds of large-scale TM projects and initiatives. Their accounts reveal not only tensions in terms of adapting to new ways of working and aligning different agendas, but also positive benefits in terms of bringing together academic and commercial knowledge and expertise. In this section, I reveal some of the conflicts and tensions that manifest when academia, industry, and public health-care providers each attempt to derive value from collaborative R&D in the life sciences. However, I also want to emphasize that people do endeavor to overcome cultural and professional differences to "muddle through" and make R&D work in practice. It is important to understand how this happens and what lessons can ultimately be learned.

The revealed professional and cultural differences between academia and the commercial research sector were perhaps the most striking in the accounts of my interview respondents. For instance, a representative from the policy community made the following observation:

> When I've talked to them [pharma] they've often said, the difference between us and academics is we want to know the clear experiments that give us confidence to invest in a development program, whereas an academic, they seem to be much more interested in a deeper understanding of how the drug is working...they will want to look at side-effects, what else is happening and why it is happening...those are important scientific questions, but from an investment point of view, they don't really guide the money guys in pharma. (Director of Policy 1, UK Public Sector Organization)

A number of senior academics supported this view, with many pointing to the different cultures of research, pharmaceutical firm scientists being milestone driven and goal oriented whereas academics prefer flexibility so that intellectual ideas can ferment. A respondent from the United Kingdom's NHS, who had worked in partnerships with both academia and industry, stated that she believed organizations like the NHS and universities had not made sufficient allowances for these very different modes of operating and institutional cultures. Nevertheless, most respondents who had worked directly in collaborative partnerships believed that conflicts and tensions tend to resolve once people actually start collaborating. Some academics admitted that industry did bring new intellectual insights, in addition to the very expensive technologies or novel compounds, which itself cannot be underestimated as a significant driver for academic involvement in such initiatives. As one clinical academic put it:

> From an academic perspective what I like is to be able to play with new toys. When I say new toys, obviously I don't mean a new piece of imaging kit or anything like that...they [Pharmaceutical company] have a tool, a toy, which is a drug that blocks a receptor or inhibits an enzyme or affects a pathway, which I never have access to unless I work with them. And then, if I have access to it, that allows me to do and ask innovative questions from a scientific perspective. If that actually in the process then answers their question for their drug TM program, then yes it is worthwhile going taking it to patients, and it's a win-win. (Senior Clinical Academic involved in a PPP)

Of course, things are not so sanguine when expectations do not align or results do not satisfy earlier expectations. For example, if an academic scientist working with a drug company discovers that the drug will not work as intended, this information has real value for the company (it can shelve that particular drug program and not waste any more money developing it). However, the academic gets very little value from this scenario, because he/she will not be able to publish the work in a top-tier journal and will derive little if any academic kudos. Career progression within universities is not generally through collaborative work with pharmaceutical companies in and of itself. Here again we see two sides to the value proposition. In this case, value to the drug company is ultimately commercial in nature. Knowledge about the compound drives a largely economic decision on product development. But value to the academic is largely scientific and intellectual in nature, and may appear incommensurable with the strict economic drivers. However, both can be realized through a collaboration, even if they appear initially to be in conflict.

Many interview respondents from the academic side did feel that there was intellectual value in working with industry and that academics could actually learn a lot from how the pharmaceutical industry operated. For example, a senior academic suggested that industry was much better at putting together interdisciplinary teams to solve problems than academics. He stated:

> They [industry] start with a problem and then work backwards, which is absolutely how you should do it. Whereas we tend not to be able to do that, but we should be able to do that much better. So you say, there is an issue...let's bring together the skills we need to solve that issue or problem, whether they be clinical modelers, systems biologists or clinicians. (Senior Clinical Academic 2)

Similarly, another respondent stated that the milestone-driven approach of industry, and the expectation that you stop a line of research or a program if a milestone is not met, was often difficult for an academic scientist to accept. Nevertheless, academia could benefit from adopting this approach. This respondent bemoaned the academic that lacks the agility to quickly stop what they are doing if it is not working and change path to explore alternatives options, which is part of the culture of industry scientists and something that is so normal and routine for them.

Since these kinds of TM-inspired partnerships require new alignments not only between academia and industry, but also health service providers, the question of how best to manage different expectations of value and benefit becomes even more crucial. In the previous chapter, I identified a number of different drivers of TM and the associated interests and values of key stakeholder communities. Industry, academia, the health services, and the policy/regulatory communities clearly have their own agendas that may not always be fully aligned, but within TM initiatives these different expectations and tensions must be resolved or at least managed. I have discussed academic/industry tensions in some detail, but the role of organizations such as health service providers throws into sharp relief the broader patient-centered outcomes. My interview accounts from health service representatives were positive about participation in PPP initiatives as a way of improving patient care, which is nicely captured in the following account:

> The health service wants anything that aids patient care, so if translational medicine leads potentially to some impact, so much the better...we've got one project that isn't going to give an immediate impact to patients, it's on imaging, but the benefit to us is that if you can ensure you're taking better images that will ultimately translate into better care. (Senior Health Services Manager 3)

Here, the health service sector must invest in a promissory future in which the fruits of investment and participation in TM will be direct improvements in patient care. Another respondent, a senior clinician from a US academic medical center, argued that it is essential that health providers be involved in these kinds of collaborations so that healthcare professionals are informed about emerging infrastructure technology and know-how, so when the technologies and tests eventually come to market they are able to engage with them effectively. They will acquire, in the nomenclature of innovation studies, sufficient "absorptive capacity." Of course, the benefits to patients and health-care providers cannot at this stage be immediate, so visions and expectations are always future oriented. This contrasts with the greater sense of immediacy within industry, and to an extent the policy community, where the economic bottom line and justification for investment decisions are paramount and there is a recognized urgency to respond to the broken middle of R&D and nurture the bioeconomy.

However, in the twenty-first-century bioeconomy, we cannot ignore the fact that health services are critical to the long-term success of TM, because of the crucial link to patients and tissue samples which have different types of value depending on how they are used, by whom, and for what purpose. One interview respondent described the heath service as the most valuable asset in any TM initiative as it is the gatekeeper to patients, tissue samples, and data linkage. However, there was a feeling among some respondents that the value and expertise of the health service are often overlooked. One described the contracting process for a particular partnership as follows:

> The NHS was fiercely in there, it was a constant battle to say what about the NHS, what about the fact that an awful lot of what you need is coming from the NHS? So we're essential partners and you can't get the patients and the data without us. (Senior Health Services Manager 1)

The NHS, or indeed any public health service provider, is perhaps unique in the collaborative process in the sense that it has not historically been subject to conventional metrics of economic value. For industry, academia, and the policy community, there is an ever-present sense of economic value underpinning TM, in addition to the less tangible benefits, whether this is linked to commercial therapy development, tradable IP, job creation, or intellectual development. But for the health service, value has always been attached primarily to patient-centered outcomes. One policy respondent argued that in the long term, health service providers may have to begin to adopt a broader notion of value. He stated:

> If you look at the NHS, it's measured on waiting times, it's measured on incidences of infection such as MRSA [Methicillin-resistant Staphylococcus aureus], and it is measured on day's productivity. It's not actually measured on income, commercial income coming in through research. I think measurement drives behavior and we need to include this metric. (Senior Policy Representative 4)

However, deriving value, and establishing metrics or criteria for success, is not always so simple, precisely because different organizations and institutions do value things in a variety of different ways. They also operate many different measurement tools and valuation strategies. For example, in the context of the CTMM, the evaluative criteria were, according to one of my interview respondents, imposed by the government and slightly unrealistic, focusing predominantly on euros saved in healthcare expenditure. This makes sense in some cases, where, for example, a cheap diagnostic test can lead to more efficient use of expensive therapies. However, if you improve diagnostic capabilities in complex diseases such as Alzheimer's, according to my respondent, this will only add to the costs if there is no effective treatment available.

In the following chapters, I explore in more detail these kinds of value-based arguments around new treatment options, but for now I use this as an illustrative example to highlight the multiplicity of value and valuation practices at play in these kinds of discussions, and which are being negotiated and traded across different professional and institutional domains. Nevertheless, the notion of value in the context of commercialization processes within health-care delivery settings, particularly publicly funded ones such as the NHS, is an interesting conundrum. How much should economic value play in national health systems, and how does it relate to other values around patient care and societal benefit?

In 15 interviews and an expert workshop conducted with NHS representatives, clinicians, policymakers, and companies in the United Kingdom in 2013, I explored the "value" of both commercial and noncommercial clinical research and how this impacted on health. There was a clear consensus that there exists a direct and positive link between the maintenance of a strong medical research system, particularly clinical research studies, and the quality of health care delivered to local patients in the clinics where this research is located. One respondent stated that there is observational data, rather than good experimental data, that health systems accrue value (in the broadest sense) by having research located there. Patients, according to the respondent, receive better care when they are taking part in research, and new and better technologies and therapies are brought into health systems faster as a result of them having both a research and teaching function. Another respondent stated that areas with significant research capacity will attract the highest-quality doctors and nurses, who will not only drive the research agenda but also deliver quality health care to patients. So the message here is that policymakers ignore the value of the link between health care and the medical and clinical research systems at their peril. However, some respondents felt that commercial research within health services was treated with suspicion and derided by many health-care professionals. One senior R&D manager in the NHS, who did value commercial studies, stated:

> There is a thought out there with consultants that commercial studies are somehow less important than academic studies, and by coming from an academic research background myself I understand that opinion...however, commercial studies will pay for other things, if you're income generating a significant amount of money from commercial studies, you can maintain a research nurse, and the research nurse is

then able to coordinate your noncommercial activities...But there is definitely a cynical attitude in the NHS towards commercial research. (NHS Senior R&D Manager 3)

Here, the respondent suggests that the very existence of economically driven commercial research engenders broader patient benefits and clinical values. If organizations like the NHS were to operate with a more structured and sophisticated approach to generating income through innovation (research and clinical trials), as is routine in many hospitals in the United States, some felt it could play an increasingly important role in driving the TM process and generating the broader patient benefits. However, Will (2011) talks about the "multiplication of value" in NHS research, but questions the appropriateness of organizations like the NHS becoming a de facto Contract Research Organization (CRO) for industry, which some fear is being pushed aggressively by a strong policy agenda. She sees the call to locate commercial studies within public health services being justified on opaque and ambiguous appeals to the future realization of numerous patient benefits and values, which rarely materialize. This more skeptical view was not generally supported by health-care practitioners and research directors actually involved in clinical trials and TM consortia, but it does again raise the possibility that some of the rhetoric around value and valorization may be based on unrealistic, future-oriented expectations and not grounded in current realities.

The organizational/institutional challenges of TM, including the management of different organizational cultures, values, and expectations, were considered by most interview respondents (both UK- and US-based) to be more significant than the scientific and technological challenges. There was general optimism that the science and technology would continue apace, and new methods, techniques, and technology would eventually overcome current challenges and become embedded in clinical practice. But the organizational challenges were considered substantial, as captured in the following account:

> The two big challenges are that you're trying to integrate people that exist across many different structural entities within an institution, and you're dealing with a large amount of money. On the other hand, the mandate that you have is extraordinary expansive. One of the biggest challenges is communication, and that is to make people aware of what was available and why they should care, and to maintain the visibility of the enterprise. Because these types of endeavours are culture changes, they're not like a 1 year or 4 year project; it's really a 10–15 year culture change. (Senior Academic Clinician 4)

These cultural differences are particularly salient in the context of IP, and I want to finish this chapter by reflecting on this important aspect of PPPs.

Valuing IP in TM Collaborations

As Bubela et al. (2012b) argue, collaborative R&D models require IP rights to be recalibrated and creative forms of governance to manage expectations and make collaboration work in practice. Some of my respondents felt that there was a fundamental difference between academic partners and industry partners in terms of how they value IP. On the one hand, there was a feeling that industry aggressively protects their potential drug targets with walls of complex IP but that academics also had unrealistic views about IP. One industry respondent stated:

> I personally think universities overrate very early stage IP and don't realize that you have to develop that IP for it to be worth anything, and the development can cost a lot... It's often very difficult to work with the UK because the universities are very precious about their IP. They certainly value it much higher than a pharma company would... Some companies would much rather work with universities in the US. They're not so precious about their early-stage IP. (Senior Executive, Pharmaceutical Company 6)

With both academia and industry having unrealistic expectations around IP, it is possible that unnecessary hurdles to the successful exploitation of drug targets are being created. If partners in PPPs are driven entirely by the monetization options generated by complex IP arrangements, and unrealistic expectations of their long-term economic value, the broader benefits of collaboration may be usurped. The precompetitive agreements of the case examples I presented earlier should, in theory if not always in practice, avoid the over-valuation or under-valuation of IP. However, in the twenty-first-century bioeconomy, it is still all too often easy to reduce R&D to the crudest of economic metrics. Interestingly, the health service partners I interviewed had perhaps the most realistic view of IP, as captured in the following statement:

> We are quite realistic, unlike other [commercial] partners in that we probably think the value of IP will be small. And that's not a concern to us. We've got one collaborative project on imaging that is not going to generate IP and won't give immediate benefit to patients, but the

benefit to us is that if we can begin to ensure we take better images that does ultimately translate into better care for patients in the future. We want anything that aids patient care. (Health Service Manager 5)

This statement captures the more distant, and perhaps ephemeral, benefits and value that can accrue from interdisciplinary research that cuts across professional and institutional boundaries. These should be recognized as just as important as the short-term economic benefits and scientific outputs. As stakeholders muddle through the mundane day-to-day practice of interdisciplinary and collaborative research, multiple types of value can be realized if different expectations are recognized and managed appropriately.

Conclusion

What I have attempted to demonstrate in both this chapter and chapter 3 is the changing face of R&D in the health bioeconomy from the perspective of the different professionals involved in these organizational and institutional practices and experiments in interdisciplinary collaboration. Case examples have shown how actors work in new organizational regimes and negotiate, trade, and attempt to realize different values and expectations. It is clear that there is a powerful rhetoric about multiple values and benefits accruing from cross-sectoral and interdisciplinary research, and also an ingrained belief that the nature of twenty-first-century biological science requires these organizational innovations and transformations of practice within both the laboratory and the clinic. Nevertheless, despite some skepticism, it is clear that policies around TM have had a material impact on the organizational structures of academic and commercial research and engendered new professional alignments and accounts of value and worth.

By looking at both macrolevel policy and organizational change, and the microlevel of R&D practices, I have revealed some of the substantive changes in the nature of contemporary R&D and how, despite some lingering challenges, R&D can be made to work in practice, potentially deliver multiple benefits, and meet diverse stakeholder expectations. Even when collaborative research does not ultimately deliver the desired economic benefits, or fails to meet the expectations of industry (as in the case of the TMRC), these endeavors may still deliver broader value to the innovation ecosystem and contribute positively to its evolution. Indeed, we should consider all of these attempts at restructuring R&D as novel experiments, just as important as the basic science, which, together, through trial and error, success and failure, may meaningfully contribute to the evolution of new biology, the innovation ecosystem, and its related bioeconomy.

In the following chapter, I turn to the broader role of regulation and policy in the context of the innovation ecosystem and new health bioeconomy, and explore how the governance of new therapies and their pathways to the clinic are being fundamentally shaped by the challenging science and technology, and continuing uncertainties about value and benefit.

Chapter 5

Regulation, Policy, and Governance of Advanced Therapies

Introduction

In this chapter, I extend the analysis beyond the innovation communities and funding agencies that have driven new biology and translational approaches to health innovation, and explore the reciprocal impact of regulation and policy on the therapeutic innovation strategies that are driving the health bioeconomy. The question is: *How has new biology both challenged and transformed conventional regulatory systems and the resilience and adaptive capabilities of health-care systems to innovative therapies?* Using the case examples of regenerative medicine (RM) and personalized/stratified medicine, I reveal the regulatory and policy challenges facing disruptive therapies that do not have established routes to market, nor conventional business models and value chains to facilitate entry to the clinic. Many of these therapies, and their underlying or companion technologies, if they are to be successful, must find a way to fit into or transform existing health-care pathways. They must also navigate complex and onerous regulatory and reimbursement systems, which have built up incrementally over many decades.

In the context of a turbulent innovation ecosystem, I suggest that clinical, scientific, regulatory, and commercial values are being recalibrated as new product development strategies are considered for non-conventional biological therapies (particularly cell therapies) or novel "bio-objects" (Holmberg et al., 2011). New treatment options challenge the status quo and raise questions about the appropriateness of conventional regulatory practices and evaluative tools, such as the preclinical animal model for safety and the "gold standard" three-phase clinical trial for therapeutic efficacy. After briefly describing the background to modern medicines regulation, I explore how the Food and Drug Administration (FDA) in the United States and the European Medicines Agency (EMA) in Europe have responded to the myriad challenges of new biology by developing novel regulatory innovations. These include "adaptive clinical trial design," "conditional approval," and the establishment of new legislation and guidelines for "advanced therapies." This section provides an overview of some of the broader changes in regulatory strategy that have co-evolved with, and ultimately shaped, new therapeutic options.

I then identify and describe some key aspects of RM and stratified medicine that test the appropriateness of the regulatory and innovation system for new therapies, as well as specific pathways to the clinic. For RM, I explore this in the context of the limitations of animal models for establishing preclinical safety, and conventional human clinical trials for efficacy. For stratified medicine, I focus on the challenges of converging therapeutic and diagnostic business models within conventional regulatory systems to deliver what many consider a valuable, patient-centered approach to health care.

Finally, I discuss the "fourth hurdle" of health technology assessment (HTA) and the perceived need for new, innovative policies and valuation practices to better support therapeutic innovation. This may be essential to deliver the broad benefits demanded by society and required to sustain the bioeconomy. Specifically, I review recent discussion around the notion of "value-based pricing" (VBP) and how this may, or may not, facilitate the successful development of breakthrough therapies such as RM. VBP is particularly interesting in that it provides, in theory, an opportunity to redefine the nature and broaden the scope of value in the way I have suggested throughout this book.

Background to the Regulation of Medicines

The highly experimental clinical science, which initiated the so-called golden age of drug discovery and development from the 1940s to the 1960s, was conducted with a degree of freedom and unbridled risk-taking that seems disquieting in our more precautionary times. Although the Federal Food, Drug, and Cosmetic Act (FFDCA) was introduced in the United States in 1938 in response to the death of over 100 people from diethylene glycol poisoning, a solvent used in elixir sulphanilamide (Rägo and Santosa, 2008), it was the thalidomide scandal in the 1960s, when children whose mothers had taken the drug during pregnancy were born with serious birth defects, which proved a watershed moment in the history of drug development and its regulation. Reflecting on the long-term significance of thalidomide, Le Fanu (2011) writes: "Missing limbs are a very prominent deformity and the pictures of thalidomide victims as they grew up over the next twenty years acquired in the public imagination a sort of symbolic significance, a metaphor of the negligence and avarice of the pharmaceutical industry" (Le Fanu, 2011: 282).

In the United Kingdom, growing concern about the safety and efficacy of drugs led to the establishment of the Committee on the Safety of Drugs in 1963, following the US 1962 Drug Amendments Act. The latter required all new drug applications to be approved on the basis of both safety and efficacy. The FFDCA, which preceded the latter amendment, only had rudimentary oversight of drug safety and no responsibility for evaluating efficacy. In the wake of thalidomide, teratogenicity (the capability to induce fetal malformation) testing for all new drugs also became mandatory in both Europe and the United States. These developments precipitated what is now recognized as modern medicines regulation. Over time, ever more complex layers of regulation have been created to ensure new therapies are safe, effective, and of good quality. The quality requirement led to the implementation of new standards for Good Manufacturing Practice (GMP) as a central pillar of regulatory science. In the United Kingdom, the Medicines and Healthcare Products Regulatory Agency (MHRA) publishes GMP guidance through the "Orange Guide," which today remains the key reference for manufacturers and distributors of medicines in Europe (MHRA, 2014). The International Conference on Harmonisation of Technical Requirements for Registration of Pharmaceuticals for Human Use (ICH), set up in 1990, also continues to drive harmonization of technical and legal requirements for manufacture and licensing of medicinal products in the United States, Europe, and Japan.[1]

Eventually, regulation included mandatory preclinical animal studies and the introduction of a three-stage clinical trial process, which became revered as the "gold standard" for testing new therapies in humans. The modern, and highly valued, randomized controlled trial (RCT) was pioneered in 1946 by the statistician Austin Bradford Hill to test the efficacy of the antibiotic streptomycin for pulmonary tuberculosis. The method quickly demonstrated its scientific value by showing clinical efficacy was enhanced when the drug was taken alongside para aminosalicylic acid. The study also provided the first clear evidence of what would become a long-term, global problem of antimicrobial resistance (Bhatt, 2010). However, the three-stage clinical trial system did not become part of established practice in pharmaceutical innovation and regulatory science until much later, when evidence of safety and efficacy became a precondition for the granting of a market authorization for new therapies. Hamburg (2010) argues that following the 1962 Drug Amendments Act, and the new requirement to demonstrate product efficacy, pharmaceutical companies used large, multistage RCTs to prove that their products worked. So it was companies that pioneered these trials and they were undertaken in the absence of a formal regulatory mandate. The situation changed in the mid-1980s, when extensive clinical studies did become a formal regulatory requirement (Woodstock and Woosley, 2008). This highlights how informal industry practices, voluntarily undertaken to build the evidence base for product efficacy claims, can, over time, become embedded within regulatory legislation.

Once new legislation and guidelines for regulatory science became firmly established as a normal part of the drug development process, there was a ratcheting up of regulatory requirements. This increased the costs of research and development (R&D) and lengthened the time taken to get a new therapeutic compound into the clinic. Le Fanu (2011) questions whether the exponential increase in data requirements for regulatory approval of new drugs since the 1960s has significantly contributed to the development of safer and more effective medicines. He provocatively asks whether, instead, it has merely served to provide the "appearance of thoroughness" in an increasingly risk-averse world.

The emergence of novel therapies based on new biology in the latter decades of the twentieth century, and unconventional diagnostic, device, and drug combination products, began to test the limits of these conventional regulatory regimes that had been designed for small-molecule drugs. The latter were well understood by both the multinational pharmaceutical industry and the agencies that regulated it. Furthermore, because more innovative and potentially pathbreaking treatments (particularly new biologicals and more recently tissue-engineered products and RM) disrupted conventional pathways to the clinic and were expensive, HTA and new governance regimes for reimbursement emerged as a significant "fourth hurdle" facing drug developers (Cohen et al., 2007; Paul and Trueman, 2001). Some authors have reported evidence that price regulation in Europe, for example, has had a negative effect on incentives to invest in drug R&D (Eger and Mahlich, 2014).

New Biology's Challenge to the Regulatory System

New biology, as I described in chapter 2, challenged, although it did not vanquish, the "business as usual" workings of the pharmaceutical industry and the blockbuster model of drug development it pioneered for relatively simple, small-molecule drugs. New technologies and therapeutic options also challenge regulatory systems, which struggle to keep pace with scientific advances and ensure marketed therapies are as safe and effective as possible. It is important to stress here that there will always be an element of risk and uncertainty with any new therapy, so regulation can never provide absolute certainty that a new chemical or biologic is safe and will work as intended. As one senior pharmaceutical executive stated: "...all drugs are basically poisons. If they interfere with your metabolism in some way that is a poisonous effect" (Senior Executive, Pharmaceutical Company 1). Another elaborated:

> There is a kind of inflated expectation that you can get great efficacy and absolute safety, but that is just not possible...for safety, there may be a range of responses because of pharmacogenetics or other individual variations because of co-medication, because the patient didn't take the drug properly or the doctor misdiagnosed. There's a whole range of reasons why things might not be safe. (Senior Executive, Pharmaceutical Company 7)

As pharmaceutical firms struggled to adapt to the life sciences in the 1990s, so did the regulatory agencies that had built up their knowledge and capabilities around conventional small-molecule drugs. The "precautionary approach" to regulation and governance replaced the risk-taking culture of medical innovation that defined the 1940s–1960s. Then, most newly discovered drugs were provided to patients with no toxicity testing or clinical trials and often within weeks of being first synthesized. This change in regulatory philosophy increased the cost and time to bring new products to market, and at times threatened to inhibit the development of innovative therapies that did not fit the conventional pharmaceutical mold (Tait, 2007). Nevertheless, innovation in regulatory science, and adaptations to the regulatory system itself, was required to bring many of the path-breaking biological therapies to patients. For example, recombinant proteins in the 1980s and 1990s, and monoclonal antibodies at the turn of the twenty-first century, required modification of established regulatory standards, guidelines, and the very constitution of regulatory decision-making committees.

Regulators in both Europe and the United States also created new pathways to market authorization, such as "fast-track" drug approval and "orphan-drug" legislation (Messner, 2008; Milne and Tait, 2009), as a means of supporting innovation in areas of unmet medical need, or where patients required early access to experimental medicines due to the lack of alternative treatment options.[2] These innovations were often driven by lobbying from patient-groups (Epstein, 1996), as I explore in the following chapter, and are a good example of how governance has been used to support breakthrough innovation. Also, as Messner (2008) argues in the context of the FDA, many of these regulatory innovations, especially fast-track legislation, built on and formalized practices that were already being used on an informal and ad hoc basis.

However, while these examples demonstrate the willingness of regulators to consider new regulatory pathways to support innovation in cases where there is a clear market failure, or delays in approval for highly valuable and much needed therapies, neither fast-track nor orphan product designation challenged the fundamental tenets of the conventional drug regulatory model. They represented piecemeal modifications to what was a fairly inflexible regulatory process. Regulators tend to add, rather than reduce, layers of regulation when faced with new technologies or therapeutic paradigms. Certainly, regulators created incentives for companies to develop particular products, with high patient and societal value, which they were unable or unwilling to do under normal regulatory and market conditions. As Milne and Tait (2009) argue:

> Orphan product program consists of both push and pull incentives that reduce the fixed costs of R&D and regulatory approval, while increasing the expectation of profits, by providing monopoly market conditions...In the aggregate these incentives provide an economic rationale that makes the orphan drug market more attractive to drug developers. (Milne and Tait, 2009: 740)

Nevertheless, this pathway did not radically change the scientific data requirements and evaluative criteria for judging safety and efficacy. The conventional clinical trial system continued to be valorized and served as the bedrock of regulatory science. Although smaller clinical trials, and some changes in statistical methods, were permitted for orphan products due to their much smaller patient populations, many of the foundational principles of conventional preclinical and clinical testing remained firmly in place.

In future, a more "proactionary approach" (Fuller and Lipinska, 2014) may be required, which recognizes calculated risk taking as a fundamental driver of human progress and crucial to the successful development of new, experimental therapies, which are often unlike

any conventional drugs or biologics and may have no regulatory prec-
edent. This is particularly relevant to RM, which I discuss later. The
question is how far are regulatory institutions able and willing to go
to enable innovation and the realization of value from new biology
and its unconventional therapies? What is the best way to maintain
high standards of safety and efficacy without constraining innovation and disrupting the very foundations of the fragile bioeconomy?
Regulators do recognize their role as both gatekeepers to protect
public health and enablers of innovation (Ehmann et al., 2013), but it
is not clear how proactive are they are in practice.

This is a growing concern for range of innovative industries,
which was recently captured in the European Risk Forum's (ERF)
"Innovation Principle Letter" sent to President Juncker of the
European Commission on November 4, 2014, and signed by 22
CEOs of a range of major companies. The letter welcomed European
initiatives to promote innovation, but also expressed concern that
the broader European regulatory environment tends to focus on risk
avoidance rather than risk management, which encourages precau-
tion. The letter also called for a new principle to ensure any change
in regulation be implemented only once the impact on innovation
has been assessed (European Risk Forum, 2014). It is within this
broader context of regulation and innovation interactions that I critically explore the regulation and governance of advanced therapies,
and reflect on how regulatory institutions are adapting, or not, to
significant change in the underlying science and technology.

FDA Responses to Advanced Therapies

In the United States, therapies based on advances in new biology
are mostly regulated by the FDA's Center for Biologics Evaluation
and Research (CBER). On its website, the FDA describes the CBER's
mission to protect and enhance public health through appropriate
regulation of cutting-edge biological products which, unlike conventional small-molecule therapies, are not easy to identify, characterize,
and manufacture.[3]

There is a long history of the regulation of biological products
in the United States. The Biologics Control Act was passed by
Congress in 1902 to control the manufacture and use of biological products, which were becoming widely used but were not subject to any meaningful regulatory oversight. These consisted mainly
of vaccines, serum, and antitoxins. Since they were manufactured in
animals using bacteriological processes, they eventually became a

public health issue. For example, an antitoxin for Diphtheria was produced at the turn of the twentieth century by inoculating horses with diphtheria bacteria, then bleeding the animals to obtain the blood serum. This serum, which contained the vital antibodies, was then injected into patients suffering diphtheria. However, in the absence of robust standards in the production process, and poor knowledge and understanding of safe and effective therapeutic dose, there was inherent risk to the patient taking this type of product. The Hygienic Laboratory of the Public Health and Marine Hospital Service, which was a predecessor to the CBER, was established to license the sale of biological products and assess their safety. In 1948, the Hygienic Laboratory became the National Institutes of Health (NIH), which was responsible for controlling biologics until 1972, when the FDA took over and set up the Bureau of Biologics. This was renamed the CBER in the late 1980s. The FFDCA and the Public Health Service Act of 1944 are the principal laws that today still govern biologicals in the United States.

There are various divisions within the CBER, and each takes responsibility for different categories of biologics-based products. Nonbiological therapies, including all conventional small-molecule drugs, are regulated by the Center for Drug Evaluation and Research (CDER). However, in 2003, some biologics were transferred from CBER to CDER, including monoclonal antibodies, many therapeutic proteins, immune-modulators, and growth factors. Nevertheless, advanced cellular and tissue products, gene therapy, vaccines, and blood products continued to be regulated through the CBER, in recognition that these are complex and advanced therapeutic products that may require a different form of governance and regulation to more conventional therapies.

The reason for transferring certain products to CDER was to consolidate and streamline review processes at the FDA and reduce duplication of activities (Holland-Moritz, 2006). It also took a selection of biological therapies that were of growing interest to the pharmaceutical industry, particularly monoclonal antibodies, to the more familiar CDER process. However, at the time of the proposed change, there was some concern from industry that CDER would impose inappropriate requirements on these biologics, such as preclinical testing protocols designed for small-molecule drugs. Still, there was general optimism among key stakeholders that the change was necessary and would have a positive impact on innovation.

The FDA's Critical Path Initiative was initiated in 2004 as a strategic response to the challenge of regulating new and emerging advanced technologies and therapies and supporting innovative drug development. It started with an important White Paper, published in 2004, titled *Innovation or Stagnation: Challenges and Opportunities on the Critical Path to New Medical Products* (FDA, 2004). This document signaled the FDA's recognition of its important role in meeting the innovation challenge. The Critical Path Initiative was focused on responding to the translational gap between basic science and clinical practice. It challenged the biomedical community to develop cross-sectoral initiatives to bring new technology and analytical tools (genomic, imaging, informatics, etc.) into the process of drug development and review (Baratt et al., 2012). Much of the White Paper focused on the need for collaborative partnerships and precollaborative research consortia to share and exchange knowledge relevant to regulatory science, which led to the launch of the Critical Path Institute in Arizona as a not-for-profit corporation to support the Critical Path Initiative (Woosley et al., 2010). This was very much in the spirit of the public-private partnership (PPP) model that I discussed in chapter 4.

The FDA's underlying philosophy in the White Paper was that safety issues must be identified as early in the drug development process as possible, which chimed with industry's desire to reduce its phase 2 attrition rates and enable more rational investment decisions early in a product's life cycle. The FDA cited product testing for contamination, as well as in vitro and animal toxicology studies, as key to establishing this safety profile. It is interesting that conventional tools of regulatory science—in this case the preclinical animal model, which is problematic in the context of many cell therapies—can remain highly valued even in a document advocating for innovative change in regulatory science.

Nevertheless, the FDA did recognize that many of the conventional methods of assessing safety and efficacy were decades old, and that there was scope for new technology to better support regulatory science. The White Paper identified the critical path as a long process from drug candidate identification all the way through to marketing, and highlighted that while major investment and progress had been made in basic research, very little had changed in the development process (Woodcock and Woosley, 2008: 4). One opportunity cited by the FDA, in response to this problem of regulatory science, was the potential use of human cell lines for characterizing drug metabolic pathways and providing a simple in vitro method for predicting human metabolism. Other important developments included the identification and validation of novel biomarkers, use of new imaging techniques, and better use of surrogate endpoints in safety and efficacy studies. All these were listed by the FDA as important areas to develop for the future.

In two follow-up reports to the White Paper—*the Critical Path Opportunities Report* (FDA, 2006) and *Advancing Regulatory Science* (FDA, 2011)—the FDA extended the discussion on regulatory reform to the streamlining of clinical trials and adaptations to their structure and organization. Both reports noted that most clinical trials are "empirical," meaning that they are designed to assess whether patients improve or experience adverse reactions to a new therapy. They have not historically been designed to explore the underlying physiological mechanisms of product performance, largely due to limitations in the knowledge base and the lack of appropriate or reliable evaluative tools. A major drawback of empirical trials, according to the FDA, is that only a few questions can be addressed in any one experiment, leaving many questions about product performance unanswered.

As new technologies and techniques emerge that can identify the causal mechanisms of drug safety and efficacy, the FDA suggests that clinical trial design should also adapt. The *Critical Path Opportunities Report* report makes reference to what are called "learning trials," which have a different conceptual framework and statistical approach to empirical trials. One illustrative example provided by the FDA is a dose or concentration-controlled trial, which exploits biomarkers or other intermediate endpoints to identify dose response relationships.

> In the future, we hope that such trials can employ multiple biomarker assays, such as advanced imaging techniques and genomic- and proteomic-based tests, to quickly reduce uncertainties around product performance. Knowledge gained from learning trials can be incorporated into quantitative computer models of disease and product performance to refine their precision and lead to more efficient *confirmatory* trials. More conceptual work needs to be done in advancing the design and analysis of these trials. (FDA, 2006: 12)

Here, new information system technologies and modeling techniques are seen as highly valuable tools for evaluating safety, efficacy, and product performance. The FDA has also supported the need to incorporate the measurement of patient responses, and even patient preferences, in clinical trials. This is very much in the spirit of the "patient-centred approach," where patient-reported outcomes become a major component of the evaluation criteria used to determine product performance and value in its broadest sense.

Although I have only provided a snapshot of some of the regulatory changes being proposed at the FDA, the point I wish to emphasize is that major regulatory reform, in theory if not always in practice, is being actively discussed by regulatory agencies in order to contribute to solving the innovation challenges in the twenty-first century bioeconomy. However, Europe has perhaps been more progressive in trying to formally implement new approaches to enable innovation of unconventional therapies, particularly through the establishment of the Advanced Therapies Medicinal Products Regulation (ATMP).

Europe and the Regulation of Advanced Therapies

The EMA, like the FDA, has for many years tried to adapt to rapid change precipitated by advances in new biology. It has implemented various regulatory initiatives, and published discussion papers, to think through how regulatory science might be better used to support innovative therapy development. The EMA's equivalent of the FDA's Critical Path Report was the *Road Map to 2015* (EMA, 2011). This document acknowledged the Agency's responsibility to both protect public health and promote innovation. A number of important regulatory innovations have been implemented in an attempt to support innovation. In addition to "accelerated assessment,"[4] which is similar to the FDA's fast track process and applies to important new therapies for unmet medical need and that promise real patient benefit and value, there have been three key additional innovations (Mittra et al., 2015).

First, there has been the development of "conditional approval"[5] and associated "exceptional circumstances licensing"[6] as a means to encourage and appropriately regulate early market access to innovative therapies. Exceptional circumstances licensing was first introduced in the early 1990s to permit approval for a new drug in circumstances where it was not possible, either for ethical reasons or because of the rarity of the condition (mainly designated orphan diseases), to provide the complete pharmaceutical and preclinical data package generally demanded by regulators. Over time, this regulatory innovation has been expanded to include therapies for more common conditions, but where there exists unmet medical need, such as acquired immune-deficiency syndrome (AIDS) (Boon et al., 2010). Conditional approval represents a specific adaptation of exceptional licensing. It applies both to products intended to prevent, treat, or diagnose seriously debilitating or life-threatening diseases, and to products that have official orphan designation, or are intended for emergency use. The latter may be in response to recognized threats to public health, as defined by the European community or World Health Organization (WHO).

Conditional approval is valid for one year on a renewable basis, but the expectation is always that any incomplete data will at some point be provided, and this requirement is what distinguishes conditional licensing from exceptional circumstances licensing. Boon et al. (2010) claim that the principal aim of conditional approval is to provide early patient access to novel therapies, and it achieves this by compressing the development phase (clinical trials) of R&D, although expediting the review process is not itself the central objective of either conditional approval or exceptional circumstances licensing. In their review of how these two regulatory innovations have worked in practice, Boon et al. suggest that they have been successful in speeding up patient access without compromising safety. Some authors believe conditional approval should now be extended to all new medicines, and not be restricted to novel therapies for unmet medical need (Ray, 2009). However, Boon et al. are more cautious about recommending expanded use of these regulatory pathways. They argue that because these are still experimental regulatory processes, they require both regulators and innovators to learn through trial and error how the legislation might be used to improve patient care and ensure negative, unintended effects do not materialize. Eichler et al. (2015) talk about this in terms of the "evidence versus access" conundrum. That is, regulators must carefully balance, or trade-off, encouraging patient access with ensuring knowledge of the benefits and harms is appropriately transmitted to patients and their physicians. Furthermore, the authors suggest that both payers and patients must often balance the: "uncertainties about the net benefits with the uncertainties about both financial costs and foregone alternative treatment opportunities" (Eichler et al., 2015: 234). So, the perceived value and benefit of expedited access to experimental treatment should always be viewed as part of a much broader and complex valuation process, where there may be other closely related values, benefits, and, crucially, lost opportunities.

Finally, there is "adaptive licensing," which represents a significant change in how data from clinical trials are collected and managed. As Eichler et al. (2012) argue, adaptive licensing is based on the recognized limitations of the conventional RCT and the binary decisions of traditional drug licensing:

> At the moment of licensing, an experimental therapy is presumptively transformed into a fully vetted, safe, efficacious therapy. By contrast,

adaptive licensing (AL) approaches are based on stepwise learning under conditions of acknowledged uncertainty, with iterative phases of data gathering and regulatory evaluation. This approach allows approval to align more closely with patient needs for timely access to new technologies and for data to inform medical decisions. (Eichler et al., 2012: 426)

From this perspective, adaptive licensing is based on a more pragmatic and realistic understanding of drug innovation and data requirements for effective decision making at different stages of therapeutic development. It is similar to what the FDA describes as "learning trials." Adaptive licensing permits flexibility in the process and allows clinical trials to be adapted over time as new real-world data emerges. In a more recent paper, Eichler et al. (2015) prefer to adopt the term "adaptive pathways" to capture the fact that licensing is merely one part of a continuum from drug development to clinical use. Orloff et al. (2009) see adaptive trials as providing a more integrative and flexible model that recognizes the value of accumulative knowledge. Allison (2012) explores the issue from an industry perspective and considers adaptive trials as a curative to the spiralling costs of R&D, because in theory such trials will use smaller patient cohorts and have flexible protocols to allow researchers to modify the parameters of the experiment once it has been initiated. This flexibility is not available under the current regulatory requirements for clinical trials.

Although the system is not yet fully in place for adaptive licensing, and whether a transformational or merely incremental version of it will be adopted by regulators is still uncertain, the EMA has launched pilot projects to test how it might work in practice. In March, 2014, the EMA, through a press release (EMA, 2014a), invited companies to participate in projects with ongoing drug programs. The Senior Medical Officer of the EMA claimed that the intention of the pilot was to use real-life drug data to gather evidence that might enable regulatory changes to better balance the need for early patient access to therapy, with the important information requirements to effectively manage benefits and risks.

All these regulatory innovations do demonstrate the willingness of regulators to consider adapting existing mechanisms to facilitate innovation, but usually only in exceptional cases. However, one of the biggest changes implemented by the EMA in response to potentially breakthrough therapies emerging from new biology was the development of a specific new regulation known as ATMP, which created a separate pathway for a defined set of advanced therapies.

The ATMP Regulation and Its Vision for the Future of Advanced Therapies

In Europe, the ATMP regulation falls within the scope of the Medicines Framework (Medicinal Products for Human Use), which provides a centralized approval process for new medicines in the European community, including what are being described as "advanced therapies." This is separate from the European Commission Tissues and Cells Framework, which includes the Tissues and Cells Directives that were implemented in 2004 to centralize governance for: (1) tissues and cells, as well as manufactured products derived from tissues and cells and (2) tissues and cells for human application (research or therapy), including stem cells for hematopoietic reconstitution (bone marrow transplants).

The beginnings of the ATMP regulation can be traced to 2005, when the European Commission published a draft regulation on advanced therapies, which included gene therapy, somatic cell therapy, and tissue engineering. The European Parliament voted to approve the regulation on April 25, 2007 (European Parliament and Council of the European Union, 2007), and the regulation was approved by all members of the European Union in 2008. There were four key measures in the proposal, all aimed at creating a specific pathway for advanced therapies in Europe, given the unique nature of these products and the gaps in existing regulation that precluded any viable route to market.

First, the ATMP created a central marketing authorization procedure, and pooling of Community expertise, for all advanced therapy products requiring a marketing/manufacturing authorization. Both autologous (patient's own cells are extracted and cultured before being transplanted back into the same patient) and allogeneic (cells cultured from a single donor that are then provided to many patients) human tissue-engineered products (hTEPs) and cell therapies fall within the regulation (Mittra et al., 2015). Faulkner (2012) argues that the advanced therapy concept "...was adopted in large part in order to align tissue engineering, for which there was as yet no EU-level regulation, with cell therapy and gene therapy, which already had been subject to pharmaceutical regulation" (Faulkner, 2012: 761). However, the term "engineered" was always quite ambiguous and contested in early discussions about the proposed regulation.[7] This has long been an issue in proposed governance frameworks for hTEPs (Kent et al., 2006). Stem cells that have been extensively manipulated or modified on an engineered process are subject to the ATMP regulation. Unmodified cells used in transplants (such as bone marrow transplants, placental, and fetal stem cell transplantation) do not fall under the new regulatory framework, as these have been performed routinely for many decades and are covered by existing regulations.

Second, a Committee for Advanced Therapies (CAT) within the EMA was created to provide technical advice and criteria for evaluating advanced therapies. CAT is responsible for developing criteria and guidelines for product evaluation, using community-wide expertise, for therapeutic products very different to conventional drugs. It has a central role in establishing, in consultation with innovators, the parameters and substantive data requirements for preclinical and clinical work.

Third, special incentives were built in to the proposal to support innovation in small and medium-sized enterprises (SMEs), which play an important role in the health innovation ecosystem as they drive much of the science and technology underpinning advanced therapies and provide vital services to large pharmaceutical companies. The EMA recognized that opaque and lengthy regulatory procedures, coupled with a lack of scientific expertise in some authorities, were making it difficult for SMEs to bring advanced therapies to market (Mittra et al., 2015). Incentives included accelerated assessment, fee reductions, and use of orphan drug legislation, where appropriate. Interestingly, some of these incentives applied only to commercial organizations, so excluded noncommercial organizations such as national health services. There was an implicit assumption that small, commercial companies would be the principal developers of these therapies, and that public sector bodies would not be looking to develop industrial-scale products. The Bloodpharma case I describe later reveals the flaw in this assumption.

Fourth, and related to the last point, the ATMP made a distinction between "hospital-based" and "commercial research," by allowing for what is known as "hospital exemption" for autologous treatments. The scope of hospital exemption is

> any advanced therapy medicinal product, as defined in Regulation (EC) NO 1394/2007, which is prepared on a non-routine basis according to specific quality standards, and used within the same Member State in a hospital under the exclusive professional responsibility of a medical practitioner, in order to comply with an individual medical prescription for a custom-made product for an individual patient. (see MHRA draft Guidance and Article 28 of Regulation EC 1394/2007)

This specific derogation was defined in such a way as to allow innovative and highly experimental treatments to continue to be developed within hospitals for patient benefit, without medical practitioners having to apply for a full market authorization and meet strict regulatory requirements. However, these products are still expected to meet the safety and quality standards set by national regulatory bodies. Hospital exemption could provide an opportunity for the development of innovative therapies, on a noncommercial basis, but issues have been raised about the very definition of "nonroutine" production and whether this permits, in practice, the development of, for example, stem cell treatments within hospitals. The hospital exemption continues to be a contentious issue, and its definitional scope and application has been largely dependent on how it has been interpreted by individual member states of the European Union (Cuende et al., 2014).

The ATMP is an innovative piece of legislation that aims to reduce the risk and uncertainties faced by manufacturers of advanced therapies, although it continues to impose a high regulatory hurdle for safety, efficacy, quality, and postmarketing surveillance. As Faulkner (2012) argues, the European Commission sought to implement the ATMP to bring together a number of advanced therapeutic products within one legislative framework. This was in recognition that these products had a set of shared characteristics that differentiates them from conventional therapeutic products; namely " . . . innovative manufacturing; scarce scientific and industrial expertise; the importance of traceability and risk management; and the primary participation of small and medium-sized enterprises (SMEs)" (Faulkner, 2012: 761). Faulkner goes on to suggest that the ATMP is future oriented in that it reflects European hopes and expectations of an emerging pathway for advanced therapies such as RM. In this sense, the EMA is an anticipatory organization attempting to construct a viable regulatory pathway for new bio-objects, in recognition of both their socioeconomic and clinical benefits and value. However, Faulkner adds the crucial caveat that while the regulation conveys "generative" expectations by formulating an enabling process for product assessment and approval, it does not give a clear definitional identity to these products (Faulkner, 2012: 766).

The ATMP has undoubtedly closed some regulatory gaps for many so-called advanced therapies, and chartered a potential route to the clinic for some path-breaking products that are emerging from advances in new biology. However, the question remains as to whether this is sufficient for innovation to flourish in this area. Are approaches such as the ATMP truly innovative from a regulatory standpoint? Or, alternatively, are they merely piecemeal proposals that create a legal route to market, but do not represent a radical departure from the conventional pharmaceutical model of regulation and the norms and practices of regulatory science that have been inculcated within regulatory institutions over many decades? I now explore this question further through two case examples, one from the field of RM and the other from the emerging field of stratified medicine.

The Challenges of RM

RM (use of stem cells, or related technologies, to encourage regeneration of tissue, cells, or organs to normal function) represents one of the most path-breaking developments to emerge from new biology. It also faces significant challenges along its entire product development pathway. RM is considered to have a broad range of potential therapeutic applications, particularly for chronic and degenerative diseases. The health innovation ecosystem is not currently conducive to the successful commercial development of RM therapies, so the sector has been slow to advance and few new therapies have made it to the clinic. RM is therefore a good example through which to explore the myriad challenges facing breakthrough therapies that are unconventional in terms of their underlying technology, business models, manufacturing challenges, and key markets. In collaboration with colleagues at the Innogen Institute, I have studied the RM field in depth through a number of different projects and related publications (Mastroeni et al., 2012; Mittra et al., 2015; Omidvar et al., 2014). I draw on some of this collaborative work to explore how RM continues to challenge the current regulatory system, even though some regulatory changes have been implemented and discussion continues about further adaptations.[8]

In 2013, the United Kingdom's House of Lords Science and Technology Committee published a report (HOL, 2013) highlighting the major barriers facing the successful translation of RM therapies into viable clinical products. Major areas of concern included intellectual property and patenting practices, manufacturing capacity and health service procurement policies, HTA and reimbursement processes, lack of innovative funding models, and regulatory uncertainty. The last two are closely related, as the regulatory uncertainty, and at times complexity, is partly responsible for the low level of investment beyond early-stage, publicly funded RM research. It also partly explains the failure of many potentially promising therapies to break through to the clinic. One interview respondent (RM Senior Scientist, United Kingdom), speaking in the context of the lack of a defined regulatory path to market, pointed out that development for RM is much more difficult and expensive than other advanced therapies. Innovators must not only think through potential solutions to unanswered questions posed by regulators, but also anticipate and formulate some of the underlying questions. Another respondent, from industry, pointed to the challenge of defining RM products for the purpose of regulation, given that many RM therapies are more akin to surgical transplants than medicinal therapies. He suggested that they are often "hybrid products," being neither simply implanted medical devices, because their regulatory path is more biologics based, nor conventional biologics, since they have this strong surgical association (Representative from a RM Company, United Kingdom).

This is where recent work on "bio-objects" and the process of "bio-objectification" is particularly illustrative. As discursive, fluid, and mobile boundary-crossing entities, bio-objects (of which a stem-cell or RM product is a good example) challenge conventional legal, ethical, social, economic, and regulatory ordering systems. In this case, it is the boundary between therapy, device, and surgical procedure that is problematized by current regulatory norms. As Holmberg et al. (2012) elaborate:

> Bio-objects are in principle, contested socio-technical objects. But they depend on the existence and manipulation of living entities that have some coherent biological form and agency.... Science seeks to stabilize and classify and deploy bio-objects in novel ways, but this can be extremely difficult, as we have seen in recent years in attempts to standardize and control the use of embryonic stem cells in cell therapies. (Holmberg et al., 2011: 741)

As regulators attempt, but inevitably struggle, to standardize processes for governing RM therapies (preclinical and clinical testing protocols, as well as postmarketing traceability and patient follow-up), they are faced with the challenge of how to define the very boundaries of these hybrid biological entities, which are unlike conventional biologics or drugs. Meltzer and Webster (2011) argue that the very concept of ATMP, which had to be invented by European regulators as it did not exist anywhere else in the world, was a means of trying to resolve the problem of how best to classify, for regulatory purposes, hTEPs that involve the reconfiguration and manipulation of cells (Meltzer and Webster, 2011: 649). The authors note that the EMA previously had 25 competing definitions for these so-called bio-objects, which made it almost impossible to construct a viable regulatory pathway that would facilitate innovation.

An illustrative example of how RM challenges our conventional drug-based regulatory system is the Bloodpharma project, which is a strategic partnership funded by the Wellcome Trust and the Scottish Funding Council to develop, on an industrial scale, cultured red blood cells from pluripotent stem cell lines for the transfusion market: initially as an orphan product for beta-thalassemia patients who are at risk of transfusion problems with conventional donor blood (Mittra et al., 2015). This project, which is a collaboration between various UK universities and the Scottish National Blood Transfusion Service, aims to develop enucleated red blood cells (normal red blood cells do not have a nucleus, but those cultured from stem cell lines do, so the initial technological challenge was to remove the nuclei from these cells) and scale up the production process to take the therapy through clinical trials and to beta-thalassemia patients.

The project, which at the time of writing is still in the preclinical phase, demonstrates how conventional animal testing requirements for safety and efficacy, and the current three-stage clinical trial process, is a challenge for the product development and business model for this kind of allogeneic RM therapy. My Innogen colleagues and I worked with the Bloodpharma team to map the product development pathway (focusing on manufacturing and regulatory aspects) and better understand the innovation ecosystem and constitutive value chains for this kind of therapy (Mittra et al., 2015). Here, I focus on the preclinical and clinical testing challenges for this and related products, and tease out the key regulatory factors that impact on the development of advanced cell therapies more generally.

The Limited Value of Preclinical Animal Testing Requirements

The Bloodpharma team and indeed all developers of RM therapies are faced very early on in the innovation process with the challenge of how to design suitable preclinical studies to demonstrate safety and efficacy of their products. The expectation, from regulators, is that all cell therapies must prove the risk of tumorgenicity of products is low through a combination of in vitro characterization of the product and animal studies. Stem cells are particularly liable to cause cells to turn cancerous as their growth is difficult to control. This is why strict traceability criteria, and patient follow-up requirements, in addition to conventional safety testing, must be met by developers. However, there are no established or well-validated animal models for cultured human red blood cell transfusion, nor many other types of RM therapies. Furthermore, there is uncertainty about what a full preclinical regulatory package for this kind of innovative product should ideally look like.

Regulators tend to adopt the position that their role is simply to review proposals for preclinical and clinical studies submitted by innovators. Indeed, my interviews and discussions with regulators suggest that they see their role as analogous to a scientific peer reviewer. One former regulator stated that the developers of these new therapies understand the science far better than the regulators, so it is ultimately their responsibility to determine the requirements for preclinical and clinical work, in light of the legislation and guidelines. Regulatory officials then simply evaluate the proposals. He elaborated:

> The legal framework is really very flexible because it is thin, it doesn't say a lot. It just gives you a few principles and that's it. The detail comes through the guidelines, and the great thing about guidelines is they're exactly what they say on the packet, they're just guidance. They're an attempt from the regulators to disseminate their general learning about products or disease indications...that provides flexibility because guidelines can be completely ignored when they're not relevant and they can be added to and adapted and changed very easily. (Former Regulator, Europe)

However, regulators can, and will, always reject proposals that they judge to be insufficiently rigorous. This is why they have encouraged open discussion with innovators, before regulatory plans are submitted, to discuss options for clinical development in light of the legislation and evolving guidelines. The above quotation suggests that the regulatory system is not particularly onerous, and is infinitely malleable to changes in technology. However, this is not always the experience of developers of innovative products. A number of my interview respondents from both industry and academia suggested that the regulatory system does not sufficiently encourage innovation, and is often sclerotic and conservative:

> There is no premium awarded for having an innovative therapy as opposed to the fifth beta blocker. Regulators should reward innovation, either through accelerated approval or by penalizing non-innovative compounds. (Senior Academic Scientist 7)

Others felt that risk aversion was making it difficult to get new therapies on to the market, even if they were considered better, when valued in terms of safety and efficacy, than products currently on the market. A good example of this is the drug Warfarin for hemophilia, which was approved before new regulatory guidelines were implemented and would unlikely be approved under current requirements due to the fact that the product significantly increases the risk of internal bleeding. However, products with a better safety profile than Warfarin have been developed, but are not available because they too do not meet the current, minimal safety requirements.

In a number of workshops and interviews with members of the Bloodpharma team, various options for preclinical animal testing that regulators could demand for the cultured blood product were discussed. Some of these could be prohibitive for developers of these kinds of therapies. For example, red blood cells could be created from the embryonic stem cell lines of other species, and tested in those representative species. Alternatively, transgenic animals could be created to test human cells. Both these options would be incredibly time consuming and expensive, to the point that the business model for the therapy could become unviable (Mittra et al., 2015). The project team considered it unlikely that regulators would insist on this level of animal work, before authorizing clinical studies, but participants believed, at the very least, regulators would expect product testing in an immunocompromised animal. However, even this raises serious questions about the overall value and validity of animal testing for human cell therapies:

> Use of a homologous animal model would not test the same medicinal product as the human stem cell-derived cultured blood and, even in the latter case, a question remains as to whether the comparison would be relevant to the safety and efficacy of the product in humans. (Mittra et al., 2015: 186)

Indeed, regulators have recognized that there is an inherent problem with testing human cells in animals. In a published discussion paper, the CAT stated: "The only relevant species for testing human cells—when all aspects including receptors, cytokines and micro-environment are considered—is the human being itself" (CAT, 2010: 197). Having said that, the CAT and the EMA have concluded that there are some safety aspects of cell therapies, particularly evaluating biodistribution and tumorigenic potential, which currently can only be done with preclinical animal models. So developers of these kinds of therapies must develop a robust and justifiable preclinical plan that has the right mix of in vitro and animal testing that will be acceptable to the regulators as an appropriate method for evaluating safety, before first human trials are authorized. This is a perfect example of how innovators and regulators must muddle through the uncertainties and lack of clearly defined protocols to establish evaluation criteria for safety and efficacy of advanced therapies.

Animal testing for safety continues to be an intrinsic feature of regulatory science, an ineradicable principle that persists despite significant advances in alternative methods, such as in silico or computer modeling for toxicity (Raunio, 2011). Furthermore, the principle of the 3Rs (replacement, reduction, and refinement of animal studies), which was first devised by Russell and Burch (1959) over 40 years ago, has been positively embraced by regulatory and scientific bodies, particularly within Europe, but also the United States. A 2007 report by the US National Research Council imagines a future where all routine toxicity testing is done with human cell lines or in vitro (Raunio, 2011: 2).

The value of animal disease models and experimentation for safety and proof of concept for new therapies has long been questioned, as many treatments show success in animals but ultimately fail in humans (van der Worp et al., 2010). Lewis et al. (2012) question when an animal model (which they describe as a bio-object) is considered "good enough" to be a useful representation. Scientists must calibrate the animal against the phenomena they are supposed to represent and standardize the process as best they can, but this will never be perfect. Davies (2010) has challenged the validity and fundamental value of mouse models as an analogue for human behavior in studies designed to test depression therapies. van Meer et al. (2015) question what contribution animal studies can make to knowledge and understanding of "biosimilars" (the biological equivalent to a generic small-molecule drug), in the context of the EMA requiring animal experiments to confirm similarity to a reference biological product. In another paper on the value of nonhuman primates for the development and testing of monoclonal antibodies, van Meer et al. (2013) provide evidence to suggest that the value of testing for safety and efficacy of these products on nonhuman primates is at the very least scientifically debatable.

However, one of the most interesting analyses of the animal model challenge is provided by Kooijman (2013). She adopts an innovation studies perspective to explore why animal studies continue to be a cornerstone of regulatory science, and persist despite their known limitations and the increasingly recognized value of alternative approaches:

> The limited value of animal studies to predict human outcomes is an incentive for pharmaceutical companies to search for methods with a

higher predictive value. The limited predictive value of animal studies results in the loss of valuable drugs and makes animal studies a slippery slope to clinical trials as it provides a false sense of safety. (Kooijman, 2013: 10)

From this perspective, it would seem that there is an obvious incentive, and clear commercial and societal value, to replace many animal studies (although some animal studies are still necessary) with improved preclinical techniques and methods, especially for human cell therapies. The reason why animal studies have not been substituted, according to Kooijman, is a function of the distributed technological innovation system within which established practices become "locked in." New innovative approaches therefore find it difficult to become institutionally embedded. She writes, "... animal studies are locked-in because they are embedded in a well aligned set of institutions [and I would add established value chains] that are taken for granted, normatively endorsed, backed up by regulatory authorities" (Kooijman, 2013: 15). In the context of conventional, blockbuster type drugs, the cost and time constraints of animal studies are perhaps not so significant for a multinational pharmaceutical company. So the large pharmaceutical firms have not sought to push for significant change in preclinical requirements. Furthermore, there is a whole industry, and entrenched value chains, built on animal testing, so it would be difficult to replace this in the absence of direct regulatory intervention. However, for smaller organizations developing RM products, such requirements can be a major constraint.

Limitations of Conventional Clinical Trials for RM

Once preclinical work has been completed, RM therapies then face the additional challenge of a three-stage clinical trial process designed for conventional small-molecule drugs. For the Bloodpharma therapy, the initial market was always going to be beta-thalassemia patients, for a number of reasons. First, it is eligible for orphan medicinal product status, with the benefits of fee reductions, extended market exclusivity, and regulatory assistance. Second, because the market is small, the team would have to manufacture a lower volume of product to meet the clinical trial requirements (Mittra et al., 2015). Manufacturing and scale-up is a major challenge for this product, as it is for almost all RM therapies, and this has been widely discussed in the literature (Ratcliffe et al., 2011; Williams, 2011). Eriksson and Webster (2008) link this challenge to that of standardizing the unknown. The stem cell, or any RM product, is not a stable biological entity. It is therefore difficult both to scale up production in the same way as conventional drugs or biologics, and implement storage and distribution strategies to successfully deliver to patients. Even for small clinical trials, the manufacturing and distribution/delivery challenge is significant and potentially prohibitive.

Although the Bloodpharma product, due to its orphan drug status, would likely be amenable to a compressed phase 1 and phase 2 trial in a small patient population, the phase 3 trial would still need to be much larger. There is uncertainty about what the data requirements would be in terms of, for example, comparative studies with conventional donor blood. Furthermore, once the product is successfully approved for the beta-thalassemia market, if the developers then sought to use the product for general transfusion, much larger and conventional clinical trials would be needed for each new market. The question is whether the conventional clinical trials that apply to more conventional drugs and biologicals are fit for purpose for products such as cultured blood, and other allogeneic RM therapies that are being developed. Are the regulatory requirements a barrier to innovation in RM?

The suggested adaptations to the clinical trial system, which I discussed earlier, could potentially create incentives for innovation in RM and enable smaller organizations to develop products and take them from proof of concept to market. However, the current ATMP regulation is perhaps not proactive enough. Maciulaitis at al. (2012) show that it is mainly small companies and academic institutions that are developing ATMPs, and they continue to struggle to get these products through the complex innovation and regulatory pathways. So far, the ATMP regulation has not significantly enabled innovation and product development in this area.

An example of a truly proactionary regulatory approach can be seen in Japan. In 2013, Japan announced that it would provide a fast-track approval process for stem cell therapies (Cyranoski, 2013), which is the first example of a country creating a new pathway specifically for RM. On its way to becoming law, the change was spurred by Japan's historically slow regulatory process and the desire to facilitate home-grown Induced Pluripotent Stem Cell (IPSC) technologies. Under Japan's Pharmaceutical Affairs Law, as it stood at the time, RM therapies must, in the same way as small-molecule drugs, undergo a conventional three-stage clinical trial process to get marketing approval from Japan's Pharmaceutical and Medical Devices Agency (Omidvar et al., 2014). The new amendments created a parallel approval channel for RM products. Instead of phased clinical trials, companies would have to demonstrate efficacy in pilot studies in a small number of patients (ten or fewer) in cases where the therapeutic benefit is potentially significant, or a few hundred patients if improvements are likely to be more incremental or marginal. According to Toshio Miyata, deputy director of the Evaluation and Licensing Division at the Pharmaceutical and Food Safety Bureau in Tokyo, if efficacy could be "surmised," the treatment could be approved for marketing and, crucially, for national insurance coverage (Cyranoski, 2013). The treatments would then be subject to postapproval surveillance for five to seven years.

The expectation in Japan is that it will be possible to approve an RM therapy within three years. However, when Japan's approach was discussed in our workshop and interviews, respondents expressed concern about how this adaptation of the fast-track process might work in practice (Omidvar et al., 2014). Some questioned reimbursement options for a therapy without known efficacy, and others wondered if in the long term this was an approach likely to both support innovation and maintain the highest standards for safety and efficacy. Nevertheless, this is a good example of a country using the regulatory system proactively in response to a specific translational barrier to the development of RM. It takes regulatory innovations such as fast-track approval in the United States, or orphan-drug legislation, to a new level by fundamentally changing the clinical trial system to enable early market access. In many respects, it is a more radical version of conditional approval.

What I have tried to highlight in this section are just a couple of regulatory issues that are pertinent to advanced therapies, and demonstrate how RM challenges our conventional systems for justifying safety and efficacy of new therapies. My next example highlights the regulatory challenge of co-developing diagnostics and therapies for stratified medicine, so takes the discussion back to the business of drug development.

The Challenges of Stratified Medicine

Stratified medicine (sometimes called "precision medicine") is the term now generally used to describe what used to be called "personalized medicine" or "pharmacogenetics." After the successful completion of the Human Genome Project, personalized medicine emerged as one of the great promissory visions of what new biology could deliver to improve health care. The idea that treatments could be tailored to the genotypes of individual patients, and that drugs could be delivered more safely and effectively by identifying genetic variations in responses to pharmaceuticals, captivated industry, regulators, and society. However, concerns were also raised about the social, ethical, and clinical implications of this emerging research area (Rothstein, 2003), and some have even questioned the underlying, "neoliberal" motives of regulatory agencies in promoting this approach (Hogarth, 2015). Hedgecoe (2004) explored the impact of pharmacogenetics on clinical practice. Through two exemplar case studies, he highlighted many of the challenges in adopting the technology in the clinic. In particular, he revealed the disconnect between the high expectations of industry and the realities faced by those in the clinic trying to deliver more personalized approaches to health care and integrate new diagnostic technologies. However, my 2005 interviews with senior representatives in the pharmaceutical industry revealed, even then, a growing skepticism about the future of personalized medicine for most drug therapies. Indeed, GlaxoSmithKline (GSK), which made some of the largest investments in pharmacogenetics research at the turn of the century, significantly reduced its work in this area just a few years later, when it recognized the scientific, economic, and societal challenges. The case for economic value had not been made, and broader social and clinical value was uncertain.

Nevertheless, in recent years, as biomarker research has gained a stronghold in both the commercial and public sectors, the concept of "stratified medicine" has emerged as a key strategy for the future of drug development, although it should be noted that the term "personalized medicine" still persists in the United States, and the FDA still tends to adopt this term (FDA, 2012). The new nomenclature reflects some of the concerns raised by the notion of individualized therapies. For industry this was not economically viable and incompatible with its existing business model. Stratified medicine, rather than targeting treatments to individual patients based on their unique genetic markers, hopes to identify broad patient subpopulations based on a variety of clinical biomarkers. The biomarker may be identified through molecular, biochemical, or imaging diagnostics, and stratification may ultimately be based on the level of response to the drug (efficacy), adverse reaction (safety), or in some cases the disease, rather than the treatment, may be the basis for stratification (Mittra and Tait, 2012). The latter has defined a great deal of modern cancer research, where multiple subtypes of the disease are being categorized by both public organizations and industry. As one senior academic explained: "Breast cancer [and indeed most cancers] is not one disease for which a blockbuster drug can be developed, but it's an ensemble of orphan diseases because the molecular signature is almost different for each patient" (Senior Academic 1).

Trusheim et al. (2007) offer a nice succinct definition of stratified medicine when they write: "In stratified medicine, a patient can be found to be similar to a cohort that has historically exhibited a differential therapeutic response using a biomarker that has been correlated to that differential response" (Trusheim et al., 2007). A number of key factors are required to successfully stratify a patient market. On the biological side, there must be sufficient disease variability, multiple targets for therapeutic intervention, and different toxicity or tolerability profiles linked to the therapy (Mittra and Tait, 2012; Trusheim et al., 2011). There must also be a variety of treatment options available as well as a clinically validated biomarker. However, Trusheim et al. (2011) argue that variable efficacy or safety within the target patient population may be necessary, but is certainly not a sufficient condition for a viable stratified medicine approach. This clinical variability must be broad enough to make it worthwhile to search for an optimal therapy. In short, clinical benefit to the patient must exceed the cost of identifying specific patient subpopulations and providing the diagnostic test.

If successful, this approach may not create the blockbuster drugs that industry has historically prioritized, but the expectation from industry is that it could lead to what we might term "niche busters." Examples of drugs that have benefited from a stratified approach include Genentech's monoclonal antibody Herceptin for breast cancer, which is only effective in women who test positive for the HER2 protein, and AstraZeneca's Iressa, a lung cancer drug that is only effective in patients with an epidermal growth factor receptor tyrosine kinase (EGFR-TK) mutation. If the patient subpopulations are sufficiently large, these may represent lucrative, niche markets for new drug therapies. Furthermore, by identifying those groups of patients that might respond, or not, to a particular therapy, or suffer adverse reactions, the stratified approach could allow for better decision making on regulatory approval. Some therapies that might currently fail to get regulatory approval, due to the lack of demonstrable efficacy or safety issues, could potentially be approved if data were provided to identify nonresponders or adverse responders through a stratification process.

So there is real interest within the pharmaceutical industry, and the policy and regulatory communities, to develop stratified medicine. The United Kingdom's Technology Strategy Board (TSB), now known as Innovate UK, has funded work to explore the science of stratified medicine and think through the viable business models needed to realize its value (TSB, 2011). It foresees the need to encourage diagnostic and therapy co-development at a very early stage of R&D. The Academy of Medical Sciences has also explored stratified medicine's significant economic and technological challenges and opportunities (Academy of Medical Sciences, 2007). However, there are numerous barriers to successful implementation:

> There are several obstacles and challenges to establishing new business models for stratified medicine successfully aligning the relevant industry sectors, regulatory regimes and healthcare delivery services...Business models will require innovation in product development and also in developing closer partnerships between different types of commercial organisation (pharmaceutical firms and diagnostic companies) and the currently diverse markets they serve. This area will also require smarter regulatory environments to facilitate co-development of a therapy and diagnostic, currently subject to quite different regulatory regimes. (Mittra and Tait, 2012: 711)

In short, the innovation ecosystem requires the existence and successful coordination of many complementary organizations (pharmaceutical and diagnostic developers), institutions (health services and HTA bodies), and regulatory regimes (both diagnostic and therapeutic) to make stratified medicine work in practice. Our research on innovation models for stratified medicine (Mittra and Tait, 2012) suggested that co-development at the preclinical stage of development, the option preferred by organizations like the TSB, was the most risky and problematic, because of the high risk of failure for the therapy at this stage of the innovation process. This would reduce any incentives for a diagnostic company to be involved. Diagnostics have very different business models, value chains, and reimbursement processes to therapies, which makes co-development incredibly difficult. For instance, in vitro diagnostics are generally reimbursed on cost rather than value, whereas therapies are reimbursed on a value-based model (Goldman et al., 2013). In a co-development strategy, diagnostic firms would need to capture a greater share of the value of the therapy/diagnostic combination to make their involvement worthwhile (Ferrara, 2007). Naylor and Cole (2010) argue that pharmaceutical and diagnostic companies will generally have different views about where the value lies in the drug/diagnostic combination, which makes a truly collaborative, benefit sharing partnership difficult to implement.

However, the regulatory challenges are perhaps the most significant, in that they influence the viability of the business model and structure the incentives to innovate. Currently, co-development of a therapy and diagnostic requires that each be approved by a different part of the regulatory system, namely the therapeutic and in vitro diagnostic regulatory pathways. This may limit the extent to which new value chains for stratified medicine can emerge and grow. They are two very different sectors, with their own innovation pathways and notional ideas of value. Regulators in both Europe and the United States are considering whether companion diagnostics and therapies should be packaged as combined products, removing this historic divide and developing a single, risk-based regulatory system as the default for all stratified medicines. However, at the current time, diagnostics, particularly in the United States, require onerous clinical trials for approval. EMA regulations may be less challenging (self-certification rather than formal trials are the norm in Europe), but many interview respondents from industry predicted that European regulation was likely to become more stringent in the future.

While patient safety and product efficacy are crucial to appropriate risk management, changes to the regulatory system, or decisions about which system is appropriate, must consider the overall impact on innovation. The costs associated with gaining regulatory approval for both a therapy and the associated clinical biomarker or diagnostic tool will require careful coordination of regulatory systems, and perhaps even joint clinical trial design. However, the relative cost and risk of engaging in such a process for a diagnostic firm is likely to exceed the potential benefits, again undermining the viability of a co-development business model (Mittra and Tait, 2012).

Regulatory agencies in both the United States and Europe are actively trying to develop new initiatives to make the development of companion diagnostics easier, and encourage the identification and validation of novel biomarkers, as I outlined earlier, but there are still major areas of regulatory uncertainty. For example, there is uncertainty about how adaptive clinical trials might be implemented. In theory, they could be designed to support a stratified medicine approach. Conditional approval could also be used to support stratified medicine. Conditional approval would mean companies could take a therapy to market before the completion of phase 3 clinical trials (albeit under restricted conditions) to ensure patients can access the treatment as quickly as possible. This also has a crucial implication for pricing and reimbursement, which could be enabling for stratified medicine. There has so far been resistance by health-care providers to pay a premium for a stratified therapy, but there is also a recognized need for both pricing flexibility and value-based reimbursement to incentivise innovation. The Cooksey review of UK health funding

(Cooksey, 2006), in considering opportunities for the development of stratified medicine, supported conditional approval as a means to allow such flexibility in pricing. At the moment, when a pharmaceutical company has a product approved by regulators, it negotiates a price with buyers (such as the NHS in the United Kingdom, and various public and private providers in the United States), which is then relatively fixed. If the company then discovered the drug worked better in a particular group of patients, or had no effect in other groups, it is difficult for the company to renegotiate the price of the drug and ensure it does not lose money from the restricted market. Under conditional approval, however, if the therapy were found to be particularly effective in a patient subpopulation during the conditional approval stage (demonstrate real patient value), when the price has not been firmly established, the company is then able to set a price that adequately reflects the revised "value" of the product (AMS, 2007). This is an example of how the regulatory approval process could be used to facilitate innovation, although it uses perceived patient value to drive up the price, which would not necessarily be to the benefit of health-care payers.

These are just a few of the regulatory and innovation challenges that continue to face innovative approaches such as stratified medicine, and also RM. However, even if these challenges could be resolved, developers of new, path-breaking therapies do face the fourth hurdle of HTA and reimbursement by health-care systems. These kinds of therapies challenge the resilience of our health-care systems and the institutional readiness for innovative health-care products.

HTA and Reimbursement

Both the RM and stratified medicine case examples described above challenge not only the regulatory system for market authorization, but also HTA and the very definition of value in health care. I touched on this issue in the case of stratified medicine. The fourth hurdle of HTA, reimbursement, and adoption in the clinic has a major impact on the business models of all therapies, but particularly new, advanced therapies. It is also a particular problem in the context of developing new antimicrobial drugs to meet the challenge of antimicrobial resistance, which my colleagues and I have explored in some depth (Tait et al., 2014).

HTA is of course necessary in the context of limited drug budgets and the need to ensure funding choices by third-party payers is rationalized. As Cohen et al. (2007) argue, therapeutic value is generally the primary consideration of payers, and this is calculated on the basis of safety, efficacy, and comparison to costs and benefits of alternative treatments. In the United Kingdom, cost-effectiveness decisions are made by the National Institute of Health and Clinical Excellence (NICE), which produces guidelines that the NHS tends to strictly follow. In making its decisions, NICE collates scientific/clinical evidence on the specific intervention, uses quality-of-life-adjusted-year (QALY) indicators, and accepts contributions/comments from patient groups, health-care professionals, and other experts or stakeholders in reaching its decisions about reimbursement. In the United States, the Centers for Medicare and Medicaid (CMS) conducts its own review of cost-effectiveness for coverage of its services and, although it has no formal role in determining reimbursement for other payers, its decisions tend to be influential (Messner and Tunis, 2012). So manufacturers must demonstrate both clinical effectiveness (i.e., does the product work and is it better than others on the market?) and cost-effectiveness (i.e., does the treatment offer value for money to health-care providers?).

To date, cost-effectiveness criteria, and subsequent reimbursement decisions, have been made using relatively crude valuation metrics. QALY indicators are often criticized for not capturing the "real value" of therapies. As I mentioned in chapter 1, some believe a much broader notion of value, which may vary considerably depending on therapeutic area, is needed to better capture patient and societal benefit (Narayan et al., 2013; Porter, 2010). This is particularly important for conditions such as dementia and depression, as Narayan et al. point out, because value in treatment may be extended to consider impact on carers and broader society, as well as the individual patient. In the context of path-breaking therapies, such as RM, and to an extent stratified medicine, getting into the clinic even once the therapy is approved by regulators is difficult. For RM, it is often difficult to demonstrate value and cost-effectiveness, over conventional drug therapies, given the inflexible QALY metrics and the fact that health systems have been built up over many decades to accommodate more conventional drug therapies and their particular pricing structures. This view is supported by Webster (2007) when he argues, beyond the technical problems facing RM, "...the field will be as equally dependent on the construction of what we can think of, by way of analogy, as a *social* scaffold that will act as a vehicle through which RM becomes more widely established" (Webster, 2007: 24).

Organizations such as NICE in the United Kingdom, and CMS in the United States, are de facto regulatory institutions that govern entry to the health-care system, which itself is often resistant to novel technologies or approaches that cannot fit into existing infrastructure and entrenched practices. Since many RM therapies are fundamentally transformative of both existing health-care systems and broader innovation pathways, they are unlikely to breakthrough into the clinic unless the pricing mechanism can be adapted. It is in this context that the notion of value-based pricing (VBM) offers some potential promise.

The UK House of Lords Report on RM highlighted the inadequacy of NICE's current evaluation process for innovative treatments on the grounds that it does not consider long-term savings that offset the high upfront costs. Part of any evaluation, according to the report, should consider that early investment in the field could "unlock other treatments with significant economic impact, both in terms of savings to the health system and increased potential work productivity" (HOL, 2013: paragraph 143). Alongside adaptive licensing and early reimbursement, VBP could, according to the report, encourage the commercial development of RM and help contribute to overcoming the current funding gap.

The VBP model takes account of additional value gains and wider health benefits, beyond the traditional QALY indicators, for innovative therapies. A recent article in the *British Medical Journal* (BMJ, 2013) asks whether VBP can work in practice, and describes two key changes to NICE's existing method for assessing cost per QALY that has been considered as possible way to introduce VBP. First, the relevant costs of disease and treatment options should be extended beyond those merely falling on the health service to include carers and other social services. Broader impacts on changes in employment, for example, should be part of the formal assessment. Second, QALYs should not be based only, or even predominantly, on the duration and quality of life, but should be weighted so as to reflect the severity of illness and patient experiences at the end of life.

The *BMJ* article did raise some concerns about the proposed changes. First, using employment as a criteria in VBP could undermine the principles of equity within the NHS. Second, there could be unintended consequences of VBP:

> Extending the cost perspective beyond the NHS will favour some diseases and treatments but disadvantage others. An effective treatment for a disease with high care requirements (such as Alzheimer's disease) or which enabled employment (such as for multiple sclerosis) would involve a lower net cost (widely defined) and hence a more favourable

cost per QALY. On the other hand, a drug that extends survival in a highly dependent state, as with many recent cancer drugs appraised by NICE, could incur a higher cost and hence a worse cost per QALY under the new rules...Value based pricing will lead to winners and losers. It will also make it more difficult for clinicians to explain why some patients are denied particular drugs. (BMJ, 2013)

Some of our workshop participants and interviewees shared this sentiment and also questioned how VBP might work in practice to the benefit of RM companies (Omidvar et al., 2014). First, as there is no real clarity on how VBP will be introduced, if at all, nor the criteria to be adopted, it is difficult to know for certain if it will enable or constrain specific technologies. Second, the cost of collating and analyzing the data to support VBP (which is likely to be complex) will ultimately fall on the developers of the therapy. How much and what type of data would be required and feasible to make the broader case for value is unclear. While large pharmaceutical firms could meet high costs and time delays to make a case in a VBP regime, this could be an additional burden for smaller companies and public sector research organizations, which are likely to be the primary innovators in fields such as RM.

The Association of the British Pharmaceutical Industry (ABPI) has also raised concerns about VBP, despite being generally supportive of the idea in principle (The Pharmaceutical Journal, 2012). In order to ensure that VBP delivers benefits for patients, the ABPI has stated there should be an agreement between the pharmaceutical industry, the NHS, and the Government to deliver unhindered access to patients and fully reward innovative companies. To achieve this, it recommended, among other things, that (1) all barriers to access and uptake of medicines be identified and removed within the NHS once VBP is agreed; (2) the current definition of "improvement and innovation" be broadened to recognize the value of longer-term incremental innovation, to the long-term benefit of patients; (3) the way the cost-effectiveness threshold is set and monitored over time be agreed between the industry and Government, and (4) the processes for undertaking VBP should be as simple and efficient as possible, with minimal bureaucracy. The second point is interesting in that it highlights the concern among large multinational pharmaceutical companies that VBP could be used to undermine the value and pricing structure of "me-too" therapies, or conventional drug therapies that provide only marginal improvement on existing treatments. A more radical definition and application of VBP would, however, have the greatest benefit for RM treatments and other innovations emerging from new biology that currently find it difficult to penetrate unyielding health-care systems and inflexible reimbursement policies.

From a US perspective, a White Paper published by the company Quintiles—*Oncology Drug Development and Value-Based Medicine* (Huber and Doyle, 2010)—highlights the issue of VBP in the oncology field and how it is developing in the United States. The report suggests that with a new emphasis on treatment value to complement proof of concept, clinical trials for cancer drugs need to be redefined to support pharmacoeconomic value and ensure new drugs meet the increasing demand for value-based treatment. This includes, for example, patient-reported outcomes to reveal perceived patient value. The authors argue that in the past cancer drugs (just like HIV drugs) rarely had to meet high pharmacoeconomic value criteria as they had "special status" due to

> their role in the acute treatment of usually an incurable disease with life expectancies measured in months to a few years. This "special status" permeated clinical practice where new drugs were very rapidly evaluated in man, sometimes with a paucity of scientific rationale. This "special status" also permeated regulatory approval processes and marketing authorization as well as reimbursement. (Huber and Doyle, 2010: 3)

In this context, cost and value was rarely an issue, but this is now changing as treatment considerations are more complex and multifactorial as new approaches bring clinically meaningful improvements to patient care. It might be the case that RM has a similar innovation trajectory. The authors of the report suggest that VBP will have significant implications for the continued and successful development of new oncology drugs.

> Value-based medicine is about more than just managing drug costs, constraints on healthcare spending, health-economic evaluation, reimbursement issues, and the rise of HTAs. Its emergence and evolution in oncology is part of a broadening debate on a host of topics, including the quality of cancer care, access to treatments, and the differing stakeholder expectations of drug therapy. The value of a treatment is already being shaped and formed through HTAs and inputs into the design of insurance policies, with risk–rewards balanced based on the healthcare providers and technologies demonstrating measurable value. (Huber and Doyle, 2010: 5)

These quotations highlight that VBP is not just an issue for public health-care systems like the NHS in the United Kingdom, but also for private insurance systems in the United States and elsewhere. They also highlight the importance of value beyond simple economic accounting and cost-containment, to quality of care and the meeting of patient expectations.

Health-care systems, and payers, around the world are now looking for different metrics of value, and valuation processes, to determine reimbursement and adoption of new technologies and therapies. By projecting value beyond the immediate, or short-term, cost-effectiveness criteria and crude pharmacoeconomic models that have driven reimbursement, VBP seems to align with the inclusive value-based approach I have been arguing for throughout this book. In particular, it highlights how valuation tools and methods can significantly shape the perceived value of a therapy and its ultimate price. RM and other advanced therapies can have incredibly high upfront costs for a short treatment course, and therefore fail to meet the current cost-effectiveness criteria (Malik, 2014). However, the long-term value they may bring to patients, their families, and broader society suggests that HTA organizations and health-care systems ought to embrace a more radical approach to VBP in order to encourage innovation and clinical uptake. As Malik (2014) argues, the potential for advanced therapies to cure, rather than merely treat, means that their ultimate value may come from these broader societal impacts (Malik, 2014: 574).

VBP, in addition to modifications of the other parts of the regulatory system, could also enable the broader benefits of investments in new biology to be realized in the health-care setting. Of course, it is important to recognize that VBP can ultimately mean different things to different people. For patients, it is about accessing therapy that has real clinical benefit and improves their quality of life. For health-care systems, it is about reducing reimbursement only to those therapies that show real added value (beyond traditional QALY metrics). However, for industry, it could also be interpreted primarily as a means to charge higher prices for therapies that can meet this higher value threshold. So this highlights once again the subjective nature of value and its relationship to drug pricing. It also highlights the fact that valuation is an inherently complex and social process, and always open to contestation.

Conclusion

In this chapter, I have highlighted the important role regulation and policy plays in driving the bioeconomy, shaping the innovation trajectory, and determining the viability of different business models for new therapies based on path-breaking new biology. I have revealed how regulators in the United States and Europe have struggled to respond to rapid change in the science and technology. Although they are now thinking creatively of ways to adapt conventional regulatory protocols to facilitate innovation, often these are incremental and piecemeal, rather than transformative in a way that will truly enable radical innovation. The case examples of RM and stratified medicine demonstrate that these kinds of approaches, which require careful coordination of different business models and value chains and involve small companies and public sector organizations without the resources of a large, multinational pharmaceutical company, continue to struggle to navigate the current regulatory systems and deliver valuable therapies that will be taken up within the clinic.

Finally, I explored the fourth hurdle of HTA and discussed recent developments with the concept of VBP, which could potentially transform the market potential of these nonconventional therapies. However, while such approaches are clearly welcome, the devil will always be in the detail. For the moment, at least, nobody truly understands how VBP will be implemented and work in practice. In the next chapter, I build on these issues and look more broadly at the notion of patient values and the role that both patients and publics play in the health innovation ecosystem and the bioeconomy. It is in this context that the issues of regulation, HTA, and therapeutic value I have introduced in this chapter are further illuminated.

Chapter 6

The Role of Patients and Publics in Health Innovation

Introduction

In the previous chapters, I described and analyzed the opportunities and challenges for therapeutic development, based on advances in new biology, from the perspectives of scientists, clinicians, industry, and the policy and regulatory communities. I also illustrated how each of these groups plays an instrumental role in the constitution of the contemporary health innovation ecosystem. Together, they are driving and shaping the wider bioeconomy and its various enactments of value. The future of this health bioeconomy, and the realization of translational benefits from fundamental, basic research, is dependent on the coordination and interdependence of these distinct stakeholder communities. However, there is another important group of actors that are integral to all aspects of health innovation, and that is the patients who are the recipients of health care and therapy. At all stages of research and development (R&D) on the "bench-to-bedside" continuum, the patient is ever present, if not physically (in terms of participating in a clinical study or providing tissue or data) then as an imagined user and ultimate beneficiary of medical research.

In the context of new biology and therapeutic advances over the past three decades, there has been a clear shift in the role of patients, and variously constituted publics, from merely passive recipients (or potential recipients) of health-care services, to active participants in the R&D process. We have entered the era of the so-called involved patient (ABPI, 2013), with health-care systems, industry, and policy communities mobilizing around the discourse and rhetoric of "patient-centered" care. This is particularly resonant in the context of rare or orphan diseases, as well as the emerging field of stratified medicine.

The increasingly routine use of "patient-reported outcomes" (PROs) in clinical trials, and the recognized need for broader public participation in biomedical research, some of which is now being discussed in the context of "big health data,"[1] is indicative of the perceived value of the patient perspective in health R&D. The ascendance of evidence-based medicine as a core principle for optimizing patient care is also dependent on this widening participatory agenda. Of course, the shift toward valuing the patient's perspective may also be aligned with the recognition that patients, and their families, need to be involved and willing in order for data and samples to be collected from them on an ethical basis. So there may be multiple reasons and diverse motives for ascribing greater value to patient involvement.

In this chapter, I address the question: *What are the implications of the changing role of patients and publics in the new health bioeconomy, and how can their expectations and values be better understood and managed?* My intention is to critically explore how patients and publics are being encouraged, and perhaps sometimes co-opted, into health-related R&D in a variety of different ways and for a number of different purposes. I will also consider the broader and long-term implications for the development of new therapies, and the ability of the scientific, clinical, regulatory, and commercial sectors to capitalize on the promise of big health data.

In the following section, I describe the changing role and status of the patient in health-care innovation, in the context of patients' experiential and subjective knowledge of illness and treatment becoming recognized as an integral part of research studies. The use of PROs in clinical research highlights the extent to which this is becoming formalized as a key component of regulatory science. In the next section, I draw on the work of Steven Epstein (2007) to discuss the recognition and value that has become attached to the notions of difference and inclusion, particularly in the context of stratified approaches to disease and treatment. This further emphasizes the extent to which the exceptionalism of the individual patient, or patient subpopulation, is becoming recognized as central to biomedical research, but it is not without its challenges. From here, I broaden the discussion to explore the growing role of patient advocacy groups, particularly in the context of rare and orphan diseases, as a political and economic force. In the next section, I critically explore how advances in genomics and life sciences have created the need for more involved public, and patient, participation in research, particularly in the context of collecting tissue samples and building large data banks. This takes participation far beyond the single, relatively bounded clinical trial. These new kinds of data repositories are crucial to the success of contemporary life science research, but also have significant social, political, and technological implications. In the final section, I reflect on the recent prioritization of big data in biomedical R&D, and the hopes and expectations of future value that this has engendered. These initiatives are becoming politically salient in the context of a broader challenge to reduce "waste" and create "efficiency" in health-care management and R&D. Here, big data is being promoted as a means to fully realize the patient and public benefits of biomedical innovation, and deliver the evidence-based medicine that policymakers and payers of health care increasingly demand.

My overall objective in this chapter is to reveal how drug innovators, payers, and health-care providers are being enjoined to relocate the patient, and his/her valuable data, tissues, and subjective knowledge and experiences, from the margins to the very center of the biomedical research enterprise. The question is what implications will this have for future drug development processes and the value and benefits that can be delivered to patients and health-care systems?

Valuing Patient Knowledge

Since the mid-1980s, there has been a transformation in how medical professionals, the pharmaceutical industry, and regulators/policymakers have valued the patient perspective. In turn, patients have, through the emergence of patient support groups and organizations lobbying for better treatment options and improved health-care provision, become far less passive and unquestioning in their engagement with both medical practitioners and broader health systems and policy processes. This reimagining of the patient's role in the health bioeconomy and innovation ecosystem has a number of different levels and subtle nuances, which are important to capture.

Caron-Flinterman et al. (2005) describe the different ways in which patients have become "active partners" in health research, with their experiential knowledge increasingly viewed as a valuable resource by both biomedical researchers and clinicians. There are two key reasons for this, according to the authors. First, the involvement of patients contributes to legitimizing research and reinforces the idea of health-related research as a "public good." The involvement of all stakeholders in the process, including patients, ensures that they are given a voice, and empowered, in decisions that directly affect them. In this sense, they are recognized, or at least constructed, as active, responsible, and autonomous individuals, which is in contrast to their role as a "sick patient" (Hallowell et al. 2015). Lewin (2015) describes patients now as "collaborative agents" in health care. Second, the patient's subjective experience of illness and treatment can improve the quality of health research and ensure that it remains relevant:

> The specific, experiential knowledge of patients emerges when patients acquire some knowledge by acquaintance through becoming familiar with their own body and illness, with care and cure and with their social context. Subsequently, patients develop some practical knowledge, mainly consisting of physical and mental coping strategies. This type of knowledge is important in daily practice. (Caron-Flinterman et al., 2005: 2577)

This kind of emergent, subjective, and practical knowledge not only is important for the wellbeing of the patient, but may directly contribute to the formulation of new hypotheses by clinical researchers, the design of new medical technologies or treatment options, and the evaluation of clinical trials, which I describe in more detail later.

The challenge is how to legitimize this practical knowledge and transform it into a valid and useful form of "scientific" knowledge. As Pols (2014) argues, patients' subjective experiences, and the practical knowledge that this engenders, is not codified in any conventional or meaningful sense. Nor is it something that resides in textbooks or is embodied in patient's heads. Instead, according to Pols: "It is part of practices, devices, and situations" (Pols, 2014: 83). Drawing on the case study of Chronic Obstructive Pulmonary Disease (COPD), Pols traces how practical patient knowledge can, under the right conditions, be transformed into knowledge medical practitioners can use to determine optimal treatment. Furthermore, she reveals how it can also be used to translate medical knowledge into something useful and directly relevant to patients' lived experiences. Pols work demonstrates how formal medical knowledge, and informal and subjective patient knowledge, can be made commensurable. This is important, because, as Pols rightly observes:

> A pitfall of studying patient knowledge *in contrast* to medical knowledge is that it may reproduce a separation between the medical sciences (the study of "nature," "disease," and knowledge) from the humanities and social sciences (the study of "culture," "illness," beliefs and meaning). In such an opposition, patient experiences drift out of the realms of what one may call knowledge. (Pols, 2014: 76–77)

The importance of these attempts to better capture the patient experience, and turn into useful knowledge, is also nicely illustrated in

Mazanderani et al's (2013) study of illness narratives, and the biographical value that this produces, as I introduced briefly in chapter 1. Drawing on secondary analysis of interviews with UK patients, the authors analyzed respondents' perception of the value of their shared illness experiences. One of the major themes identified by the authors was that: "in order for value to be generated from illness narratives they had to be based on the actual experiences of real people" (Mazanderani et al., 2013: 896). Furthermore, patients recognize such narratives, and the knowledge claims they make, as distinct from "disembodied" medical knowledge. So authenticity is central to the commodity value of these shared illness narratives, and the knowledge generated from patients' experiences must also be seen as something different to other, more conventional forms of clinical knowledge. Of course, the process of demarcating patient experiences and the understandings of medical professionals on epistemic grounds should not lead to patient knowledge being considered subordinate. As Mazanderani (2014) rightly argues, patient knowledge should be acknowledged and valued in its own right. This can then be incorporated into health-care decision making for the purpose of creating better outcomes.

The knowledge patients have about their own illness and treatment does not just have biographical value for other patients, and specific clinical value in the hospital setting. It also has more tangible commercial value in an increasingly monetized global health-care context. For instance, Adams (2011) highlights how the use of crowdsourcing by businesses to generate consumer reviews and ratings is now increasingly being applied in the context of health. Here, what the author labels the "reflexive patient" is seen as an increasingly valuable resource.

> Organizations use the web to solicit and publicize narratives of individual patient experiences with health services. Patients report their experiences with institutions, professionals, medication, or treatments and may be asked to rate aspects of care...This information is available online for other patients, but is also repackaged by the site into reports that may be shared with hospitals, insurance companies, professionals, policy makers or others. (Adams, 2011:1069)

The organization "Patientslikeme," an online patient information aggregator, exemplifies this kind of approach. It is one of the largest online companies collating subjective patient experiences in order to contribute to pharmaceutical research. Adams is critical of the assumption, which often underpins the marketing strategies of these health websites, that such sites increase participation and transparency in health care. Instead, she points to the powerful role and influence of website administrators who mediate patient responses and act on behalf of a so-called "patient collective" which, according to Adams, is not predefined but actively created through the sites. In this sense, the sites have an important performative function. Nevertheless, this particular enactment of patient participation, and the recognized value of their personal experiences and subjective accounts, is part of a continuum. At another point on this continuum is the more formalized use of PROs, which are becoming an integral part of the evaluation criteria for clinical trials. They are being embraced by both commercial pharmaceutical companies and regulatory institutions.

PROs and the Validation of Patient's Experiential Knowledge

The underlying driver for using PROs has been this recognition that understanding treatment and health outcomes from the patient's perspective can improve the design of new health technologies, therapies, and treatment regimens. PROs can also contribute to improved evaluations of safety and efficacy, which is particularly pertinent to the comparative assessment of different treatment options. This is important in the context of health technology assessment (HTA) and may also have significant implications for value-based pricing (VBP), as I discussed at length in the previous chapter. If the collection and analysis of PROs can lead to the identification of subtle differences between treatment options, which cannot be captured through conventional tools, this may help shift the value of particular products and determine their reimbursement. Eichler et al. (2015) suggest that the increasing use, by regulatory and HTA bodies, of patient preferences, value judgments, and views on risk benefit of treatment is a positive development. Indeed, the authors suggest that patient representatives do not tend to advocate for early access, at all costs, but express quite balanced and nuanced views about risk, uncertainty, and acceptability of treatment (Eichler et al., 2015: 242). However, others suggest there is evidence that patients do sometimes overestimate the benefits of new treatments. Korenstein (2015) cautions that patient understanding of evidence about benefit and harm will determine the extent to which inclusion of patient perceptions in decision making improves the overall value of care:

Patients cannot properly align health care decisions with their own values and priorities unless they truly understand the likelihood of different outcomes. Patient misunderstanding of benefits and harms may create antagonism in the physician-patient relationship as patients fight for care they believe to be beneficial (and not harmful) and physicians fight back to limit harmful care or care that may conflict with the patient's own priorities. (Korenstein, 2015: 287)

So, there are clearly both opportunities and challenges in integrating patient preferences into clinical decision making, and this is now recognized by regulatory bodies.

In a recent reflection paper on the use of PROs in oncology, the European Medicines Agency (EMA) acknowledges the value of these new methodological tools for risk governance and regulation (EMA, 2014b). The EMA begins by distinguishing PROs and health-related quality of life (HRQL). The former is a broad term that includes any outcome that is based on the patient's own subjective experience and evaluation of their response to treatment. HRQL is a type of PRO, but limited to the patient's perception of the impact that the disease and/or treatment is having on his/her daily life and wellbeing. To summarize, the EMA states:

PRO is an umbrella term for the capturing of health status, symptoms, HRQL, adherence to treatment, satisfaction with treatment, etc with the emphasis placed upon the patient's judgment. It is recognized that such data are subjective, change over time and are influenced by the treatment, the disease and other co-morbidities. HRQL is a concept referring to the effect of an illness and its therapy upon a patient's physical, psychological and social wellbeing, as perceived by the patient themselves. (EMA, 2014b: 2)

So PROs and HRQL are recognized processes for capturing patient perceptions and formulating them in such a way that they can make valid clinical contributions. The data may be collected through a variety of methods, including self-reported questionnaires and interviews, but the important thing is that only patients' unmediated responses are captured. This means that there is a need to minimize bias and influence that may affect how and what patients report. In the context of clinical trial design, the EMA states that there is no standard approach for using PROs. Therefore, it is important to ensure that the underlying science and methods used are always well justified and applied correctly, and that expectations are realistic. The added value of PROs, according to the EMA, is that clinically meaningful treatment effects may be detected. This is particularly relevant where the benefit-risk profile of a therapy may be modified based on a specific range of patient responses and declared preferences.

However, there are many challenges to the PROs approach. In 2009, the US Food and Drug Administration (FDA) published revised guidance on PROs (FDA, 2009). It clarified the benefits of PROs, mirroring those reported by the EMA, on the grounds that they can measure both treatment risks and benefits, and have obvious value in both a regulatory and reimbursement context. Nevertheless, in a response to the FDA guidance, Speight (2010), a health psychology consultant, stated that there are dangers in using PROs to modify reimbursement, and that commercial drives to include PROs in drug labels are fraught with risk. In summary, PROs will only bring the desired benefits to patients, and improve drug regulatory processes, if the outcome measures chosen are relevant and interpreted correctly, clarified early in the design of a clinical trial, and related directly to prior hypotheses about treatment outcome. Furthermore, there must always be sufficient evidence to support the inclusion of PROs. Speight proceeds to note that the number of PRO-based claims on drug labels has been relatively low since the FDA first published draft guidance in 2006, which suggests that the pharmaceutical industry is finding it a significant challenge to demonstrate real treatment benefits from patients' reported accounts of their experiences and preferences (Speight, 2010: 2).

Despite the numerous challenges faced by the PRO approach, the fact that it is being given such a high and visible priority, in the expectation that it will become a routine and normal feature of drug development and clinical trials, does legitimize the claims of those who have long argued for the inclusion of the patient's voice in formal drug development and regulatory processes. This now extends beyond evaluation of clinical trials, but also to the development of diagnostic and monitoring technologies. Oudshoorn (2011), in her analysis of telecare technologies, argues that patients are now often enrolled as "diagnostic agents" of their own bodies, thus extending the clinical gaze beyond the core medical profession. "Patients are expected not only to use a technical device but also diagnose and monitor heart diseases in the absence, of healthcare professionals" (Oudshoorn, 2011: 194). This makes "patient work" an important, difficult, and highly consequential part of health innovation. Through the use of these monitoring devices, patients are also now producers of valuable biodata, which I discuss in more depth later. For now, I now want to move on and discuss some of the broader drivers for this patient-centered approach.

Inclusion and Difference as a Driver of
Patient-Centered Care

The development of personalized or stratified medicine, and indeed many aspects of regenerative medicine, which I discussed extensively in the previous chapter, narrows the clinical gaze to patient sub-populations or individuals, rather than the broad, undifferentiated, blockbuster drug markets that conventional pharmaceutical companies historically targeted. However, the transition toward greater inclusion of specific groups of patients, within which the experiential knowledge and participation of the patient have been ascribed value and recognized as a vital part of the innovation process, has a long and complex history. In his 2007 book, *Inclusion: The Politics of Difference in Medical Research*, Steven Epstein traces the broad political agendas, and policy, regulatory, and industry machinations that began to focus on diversity and equality, which led health research to begin recognizing and valuing difference. Sex, gender, ethnicity, and age all came to be seen as important variables that determine response to drug treatment and long-term health outcomes. This movement challenged an antecedent view, which had a material effect on the practice of medical research, that "white men" were the standard bearers of the human subject. Epstein writes:

> I call this set of changes in research policies, ideologies, and practices, and the accompanying creation of bureaucratic offices, procedures, and monitoring systems, the "inclusion-and-difference paradigm." The name reflects two substantive goals: the inclusion of members of various groups generally considered to have been underrepresented previously as subjects in clinical studies; and the measurement, within those studies, of differences across groups with regard to treatment effects, disease progression, or biological processes. (Epstein, 2007: 6)

The question posed by Epstein, in light of this important historical shift, is whether the categories chosen in the name of inclusion and diversity are the correct ones. Are they the most relevant for making crucial decisions on medical treatment? Furthermore, Epstein alerts us to the importance of ensuring that these ever more finely grained categories to stratify patient populations, and the sociopolitical and cultural assumptions underpinning them, do not inadvertently increase disadvantage, discrimination, and stigma. There is always a particular risk for those who fall between the gaps of any categorization system, and its changing standards, as Bowker and Starr (1999) nicely illustrated in their book *Sorting Things Out*. Redesigning how clinical trials are organized so that they include diverse social and ethnic groups, who may respond differently to treatment, is clearly aligned both to an equality/inclusion agenda and the need to improve treatment. However, it might also have unintended consequences for the broader politics of health and its shifting sociocultural idioms.

There is also the danger of creating crude categories, stratification processes, and inclusion criteria that do not align with individuals' self-assigned identities, with potentially serious implications for health outcomes. Aspinall (2013) provides a good example in his study of antenatal genetic screening for sickle-cell and thalassemia in the United Kingdom. Here, the National Health Service (NHS) used self-reported questionnaires for racial and ethnic origin in various regions within the United Kingdom as a means to identify high-risk couples (who may carry the deleterious genes and pass them on to their offspring) in areas of traditionally low prevalence for the disease. The problem, according to Aspinall, was that the preselected categories in the questionnaires did not always match individual's own, self-assigned identities. Turning this back to Epstein, his book reveals how biomedicine does not simply reflect value-neutral ideas about ethnicity, race, sex, and gender. Instead, these categories, identities, and differences are actively worked out in the everyday practice of biomedicine (Epstein, 2007: 16).

The inclusion-difference paradigm also has practical implications for how medical research is conducted. For example, Epstein points to the challenge of recruiting patients to clinical studies in the context of having to meet the requisite inclusion criteria. He writes, "Not only must researchers find willing subjects, not only must those subjects be diverse, but the groups which researchers must present include those, such as African-Americans, that routinely are considered among the most difficult of all to convince to participate" (Epstein, 2007: 15). The recognized clinical need to include a diverse range of participants in clinical trials must overcome the practical constraints of recruitment, which is a continuing challenge today in the context of stratified medicine.

Overall, Epstein is quite skeptical that this inclusion agenda, and recognition of difference, will have a broad and positive impact on health and treatment. Nevertheless, the drive for greater inclusion has been embraced by regulators and policymakers over the past three decades. It started in the late 1980s and early 1990s, when organizations like the NIH and FDA instructed drug manufacturers that they needed to start providing data on patient subpopulations, reflecting gender, age, race, and ethnicity. In 2005, the very first drug with a label based on ethnicity was approved by the FDA. The controversial drug was named BiDil, for the treatment of congestive heart failure, and was only to be prescribed to self-identified African-Americans. BiDil was not a truly pharmacogenetic drug, because it is a combination therapy (a fixed dose of isosorbide dinitrate and hydralazine hydrochloride) that was only tested in African-Americans for, arguably, commercial rather than scientific reasons (Brody and Hunt, 2006; Tutton et al., 2008). The combination therapy was never tested in non-African American populations. Nevertheless, it serves as an example of this shift of emphasis toward patient subpopulations when evaluating drug safety and efficacy. Although race was a crude and unscientific method for stratification in the case of BiDil, the potential for stratified medicine based on molecular biomarkers is now becoming a practical reality, rather than merely a promissory hope.

The trend toward increasing inclusion and identifying difference, where it is relevant, is continuing apace in medical research. Furthermore, with the increasing emphasis on orphan diseases, and recognition that many common diseases, such as cancer, can be stratified into multiple genomic subtypes, the important role of patients, and the complex politics of small numbers in clinical research, is now much more salient. I now want to broaden the discussion to think about the role of patient advocacy groups and coalitions as a significant driver of modern biomedical innovation and this inclusion agenda.

Patient Advocacy and the Politics of Drug Innovation

Many of the changes in the way that patients have been viewed and valued by the medical profession, and their increasing role and influence across the innovation ecosystem, have been precipitated by politically active patient groups. New and innovative therapies undoubtedly play a prominent role in national and international health-care management. Patient organizations are crucial in lobbying for access to better medicines, raising concerns about safety and efficacy, strategically aligning with industry and regulators to facilitate innovation, particularly in areas of unmet medical need, and directly funding research to benefit members. Webster (2007) argues that patient groups see their knowledge as based on a form of expertise rooted in a shared experience and, crucially, they: "link this experiential epistemology to a collective moral position" (Webster, 2007: 148). Patient organizations have been instrumental in helping to put rare and orphan diseases on the agenda of political, regulatory, and commercial organizations and institutions, but patient involvement is not limited to these areas.

The first case that truly demonstrated the significant political impact patients could have on drug development, and which entrenched patient advocacy within the health innovation ecosystem, was the lobbying by acquired immune-deficiency syndrome (AIDS) activists for access to new antiretroviral drugs in the 1980s. Epstein's (1996) groundbreaking analysis of this movement showed how activists not only lobbied for access to new experimental drugs, which were taking time to move through the conventional clinical trial system, but also developed expertise and understanding of both the underlying science and regulatory processes such that they could challenge conventional practices. Perhaps the most significant influence they had in this regard was redefining the nature of clinical trials by introducing the use of surrogate endpoints (in this case CD 4 count[2]) to determine drug efficacy. They also helped precipitate the formalization of fast-track or accelerated approval for life-saving drugs. Epstein highlights the importance of this movement when he writes:

> Drug regulation is one area where the sheer effect of activism would be hard to dispute. Activists were not the only ones calling for change in the FDA, but they were the key players in pushing for the approval of AIDS drugs at an earlier stage in the drug development pipeline. And although some procedures allowing early access to experimental therapies were already on the FDA's books, others, such as expanded access and accelerated approval, are new to medicine as a result of AIDS, and have resulted in the provision of such therapies to much larger groups of patients than had been the case in the past. (Epstein, 1996: 339)

The legacy of AIDS activism has been this growing influence of patient groups and disease charities in all aspects of drug innovation, regulation, reimbursement, and access. The issue of access was a particularly salient issue in the United Kingdom in 2005, when breast cancer charities lobbied for extended access to the drug Herceptin. Herceptin was a recombinant, DNA-derived and fully humanized monoclonal antibody that was approved for the treatment of metastatic breast cancer and made available to these patients in 2002. However, in 2005, various cancer charities and support groups lobbied for the drug to be made available to patients with early-stage breast cancer, even though it was not yet licensed for this particular group and cost-effectiveness evaluation had not been completed by the United Kingdom's National Institute for Health and Clinical Excellence (NICE). The upshot of this lobbying activity was that the UK government stepped in to make the drug available to all patients with breast cancer who had the relevant marker, in essence bypassing conventional regulatory and policy norms. This was also at a time when the full clinical efficacy of the drug was uncertain, so extended access to the drug was being largely determined by a mixture of media hype and active political lobbying from breast cancer charities. So this is an example of the collective political power patient groups can muster for the benefit of their members, which was reflected in the high media coverage of this story.[3] However, as Meadowcroft (2008) points out in a paper on the politicization of health and the role of patients, there is considerable variation in the power and influence of patient groups and disease charities, with the result that some diseases become more politically salient than others.

In the context of rare, generic diseases, the growing role and influence of patient support groups is perhaps even clearer. A good example is the formation of patient-active coalitions around Pachyonychia Congenita (PC) and Epidermolysis Bullosa (EB). These are two rare, dermatological genetic diseases. PC is an autosomal dominant skin disorder, resulting from the mutation of one of five keratin genes that causes malformed keratin filaments. Keratins are proteins that form filament to support and strengthen skin cells. Although not life threatening, sufferers of PC suffer constant pain and skin damage. EB is a group of genetic disorders, which cause extreme blistering of the skin. EB can be significantly life limiting. EB sufferers are supported and represented by the organization DEBRA, while PC sufferers are supported by the Pachyonychia Congenita Project. Both of these organizations have developed broad networks of shared knowledge for patient support and sought to drive medical innovation through active participation.

There are two particularly interesting aspects to the patient networks that have evolved around these two diseases. First, the PC and EB communities are exemplar case examples of international, patient engagement. One reason for this is that the very nature of rare diseases requires international engagement to generate and share the data (genetic data, case histories, and patient experiences) required to develop new testing procedures and a potential therapy. Since the diseases are so rare, the overall patient numbers are small and geographically dispersed. Global participation is therefore essential. There is currently a successful international network for patient support, including a research register, of people with PC. This is an unusual network in that it aligns patients, families, clinicians, researchers, community health professionals, and also companies and regulators in the integration of genotype and phenotype data and experiential data from patients. Second, The PC and EB communities also highlight the need for "generational engagement" and the particular challenges in achieving this. For example, there are dedicated groups and methods of engagement specifically for the teenage members of both these patient organizations. PC and EB have profound effects on the lives of the young people who are predominantly affected by these diseases, and their interests may not always align with older sufferers. PC has established a network of young people to support each other, share experiences and knowledge, which is highly relevant to both general life course issues and the development of new clinical trials and treatments. These are just two examples of the now many hundreds of rare disease groups that are actively seeking to shape the innovation ecosystem for the development of therapies for diseases that have historically been neglected.

Boon and Broekgaarden (2010), in exploring the role of advocacy organizations for neuromuscular disease research in the Netherlands, outline a number of benefits of patient engagement in research. They argue that patient organizations, especially for rare diseases, can help overcome market failure, facilitate useful knowledge exchange between users and professionals, help expedite the innovation process, play a part in mediating social and ethical debates, and increase the democratic value of therapeutic innovation (Boon and Broekgaarden, 2010: 149). Similarly, Mavris and Le Cam (2012), in discussing the results of a survey conducted by the European Organisation for Rare Diseases (EURORDIS) on the contributions patient organizations make to rare disease research, suggest that empowerment for patients depends on them becoming active partners in research, not merely participants or research subjects. Illustrative of this, of the 309 organizations that responded to the EURORDIS survey (out of 772 rare disease patient groups invited) 37 percent had directly funded research in the previous five years (Mavris and Le Cam, 2012: 239).

This significant and direct engagement from patient organizations in the research process has helped shape R&D for many therapeutic areas and created opportunities to capitalize on the expectations and promise of advanced life science-based therapy. Patient organizations are now institutionally embedded in many regulatory and policy processes. For example, the EMA's Committee for Orphan Medicinal Products, which was established in 2000, includes patients as permanent, full members with equal voting rights (Mavris and Le Cam, 2012: 242). It is no longer unusual to see patients represented at the senior levels of decision making and advisory committees, particularly in the regulatory and policy context. In this sense, patients have acquired significant political and economic power, and routinely use this to meet strategic goals for their members. In my own interviews with pharmaceutical companies, many respondents emphasized the important role of patient organizations. One stated:

> In many of the diseases that we work in, but particularly orphan diseases...there is an increasing role for patient organizations...In all these areas where we are active we have close contact with the patients and its custom now to take the patient organizations to, for example, the EMA [European Medicine's Agency], or have them be involved in orphan drug applications...So I would say patient organizations are very active now, both on the regulatory side and the development side, even investing in companies that develop drugs. (Senior Pharmaceutical Executive, Company 5)

This quotation speaks directly to the important role patient groups now play within various parts of the innovation ecosystem and their embeddedness in the R&D process.

A number of authors have traced the emergence and influence of patient organizations since the 1980s, and revealed a significant rise since the mapping of the human genome and the promises and expectations that formed around genomic medicine. Novas (2006) explores how patients, as well as their family and carers, constitute part of a broader political economy of hope. Within this political economy, people affected by serious, and often rare, genetic conditions have become recognized authorities and experts who contribute to the production of new biological knowledge and biovalue (Novas, 2006: 290). In this sense, they are indelibly connected, through their associations and activities, to the broader health bioeconomy. Novas argues that "...the activism of patients' associations that is organized around the hope of developing cures or treatments significantly contributes to the transformation of the field of contemporary biopolitics" (Novas, 2006: 290). This aligns with the work of Callon and Rabeharisoa (2008), who describe and analyze the French Association of Neuromuscular Disease Patients as an "emergent concerned group." This group, according to the authors, underwent a transformation from "passive exclusion" to "active inclusion" in the biomedical enterprise by generating the funds, through an annual telethon, to begin directly funding research.

Callon and Rabeharisoa argue that the way economic markets are structured and have evolved, coupled with advances in science and technology, has created the necessary environment and conditions for these kind of emergent groups (Callon and Rabeharisoa, 2008: 232). Such groups are challenging, according to Novas, the contexts in which biomedical research takes place; the specific modalities through which they contribute to the transformation of blood, tissue and DNA into resources for the production of biovalue; and how they contribute to the elaboration of norms relating to how biomedical research should be conducted and how its therapeutic and economic benefits should be socially distributed. (Novas, 2006: 293)

So patient groups not only participate in research as subjects, and donate their biological material in the hope of future treatment, according to Novas. They also help to organize, coordinate, and shape the R&D landscape, as these many examples demonstrate. Indeed, Rabeharisoa et al. (2014) have recently coined the term "evidence based activism" to capture the diverse set of patient organizations, users, and activist groups, as well as the different forms of knowledge engagement that they are enrolled in. Moreira et al. (2014) apply this concept to explore the ways in which dementia care organizations mobilize knowledge (both formal and informal) and deploy a range of political initiatives to make dementia care a "matter of concern."

In the context of research in regenerative medicine, where the gap between the promise and reality of therapy is narrowing but still significant, the role of patients and patient associations as activists has also been important. Chen and Gottweis (2013), in their interviews with patients who traveled to China for experimental stem cell therapy provided by the company Beike Biotech, revealed that these individuals were aware of the potential risks of the therapy, but also had an acute sense of the risk-benefit calculus. Reflecting on their data, the authors suggest that patients are citizens with reasonable expectations that regulatory authorities will protect their health, but they also now perceive themselves to be consumers of medical technology and active participants in research, even highly experimental research (Chen and Gottweiss, 2013: 203). Langstrup (2011) also reveals how patients come to project themselves as future users of stem cell technologies and therapies, when the science is still at an early stage and prospective futures are ambiguous and ephemeral. Using the philosopher Louis Althusser's concept of "interpellation," she suggests that research policy "enacts science as a series of entrepreneurial projects" (Langstrup, 2011: 574), which gives it sense of "project-ness" and direction that patients can relate to and begin to envisage themselves as a target market.

However, it is important to recognize that it is not just the patient organizations that have, independently and autonomously, reimagined the role of the patient in health-care innovation. As Bower

(2005) illustrates, the biotechnology industry was partly responsible for the emergence of, what she calls, the "patient active paradigm" from a prior rhetoric of hope. By analyzing the annual reports and marketing strategies of many of the biotechnology and pharmaceutical companies from 1990 to 2002, Bower reveals how these firms repositioned themselves in relation to their patients in order to capitalize on inchoate support for new genetic technologies for health:

> In the emerging rhetoric of the 'patient-active paradigm' patients are not only more powerful, they also bear some responsibility for their health. A feature of this model is its acceptance that patients today are on a medical continuum from conception in which health is constantly determined by a composite of genetic and environmental factors. Although there may not be any observable symptoms of illness, individuals are expected to act to optimize their health in the long term. The implicit assumption is that with complete knowledge of the genetics and environmental experience of the individual, a perfect healthcare regime can be devised. (Bower, 2005: 199)

This sentiment was also captured in many of the interviews I conducted with large pharmaceutical firms. Nevertheless, this optimistic vision of the future of therapy and the role of patients in contributing to the commercial innovation process has not gone uncriticized. There has long been concern about the alignment of patient organizations with health-care companies. Herxheimer (2003) warns that an arms-length relationship between patient organizations and pharmaceutical companies is needed to ensure that patient groups retain autonomy and the relationship always remains transparent. The risk is that patients, and their representative organizations, may be used instrumentally by commercial organizations as a means to legitimize their business activities. Similarly, a UK House of Commons Health Committee Report exploring the broader influence of the pharmaceutical industry (House of Commons, 2005) suggested that there was growing concern about the independence of many patient groups as a result of potential undue influence from pharmaceutical companies seeking to exploit patient organizations for marketing purposes. It may be difficult for patient groups or charities with little funding to decline financial support, and their interests can all too easily become aligned with a specific company's short-term commercial strategy. For example, my interview respondent, who talked about how pharmaceutical companies now routinely take patient groups to meetings with regulators, raises questions about the independence of these groups. Nevertheless, despite ethical concerns about autonomy, process, and influence, there is a consensus that patient group involvement in health R&D has, in general, been a positive development and something that should continue to be embraced.

So far, I have focused on the role of individual patients, and now representative patient organizations, within the changing health innovation ecosystem. I now want to look more specifically at the broader role of patients, and publics, in interdisciplinary life science research that requires access to large data sets and personal medical information. This then leads onto the final part of this chapter, where I reflect on the more recent emphasis on the opportunities and challenges of big data in health care.

Biobanking and Data Linkage

Since the successful mapping of the human genome, and the emergence of numerous therapeutic options from new biology, the requirements for tissue and data collection, storage, and use have exponentially increased. Since the 1990s, the biobank has become a familiar, and often controversial, asset in the health innovation ecosystem. A biobank is a specific type of biorepository to store human biological samples (tissues, cells, data, etc.) for use in research and potentially for therapy. Donations may be made from patients or healthy volunteers, and such repositories are considered crucial for the further development of genomics and life sciences research and their translation into therapy.

The ability to identify and track gene-disease relationships depends on advanced sequencing technologies and the ability to collect and curate multiple samples. The more recent promise of Genome Wide Association Studies (GWAS), which attempts to track mutations across multiple genes and populations in order to identify those associated with specific disease phenotypes, requires this continuing capability to build diverse data banks and bridge the biological and clinical divide through advances in bioinformatics (Sarkar, 2013: 192).

There are now many types of biobanks and they vary in terms of their scale and scope, and whether they are public, private, or mixture of both. As Rial-Sebbag and Cambon-Thomsen (2012) note, the collection of biological materials and generation of biodata initially emerged in the context of human clinical trials, often as a byproduct of clinical studies. Regulatory frameworks and governance mechanisms were therefore designed for research, and its biological artifacts, relating directly to individual patients. However, the contemporary biobank concept, according to the authors, challenges these conventional norms and practices:

> Biological samples are not persons but human materials; biological samples without data attached to them are not very useful for research; the research time frame for projects using biobanks is much longer than that of any individual project, and multiple uses of biological samples over years has become the expected destiny of biobank samples. (Rial-Sebbag and Cambon-Thomsen, 2012: 113)

From this perspective, biobanks represent a step change in the nature of patient and public participation in R&D and require appropriate regulation and governance that takes account of their distinctive social, ethical, and technological features and idiosyncrasies.

Many authors have discussed the social, ethical, and legal implication of biobanks, as well as the general biobanking process. Much of the literature has focused on consent procedures, ownership, and privacy in a legalistic context (McMahon and Harmon, 2012). Shickle (2006) raises the "problem of consent" in these prospective biobanks, because the participants, at the time of first recruitment, will not know what future disease they may be susceptible to. Furthermore, researchers will not know what they may want to do with the samples or data at some unspecified future date. For this reason, "open consent," without the requirement to go back to donors for reconsent when the material or data is used for a different purpose, has tended to be the preferred option for developers of biobanks. However, this raises the question of whether consent under these conditions is sufficiently "informed." There may indeed be a tension between what is ethical for the participant and what is practical and feasible for the biobank. However, in some cases, the messages can be confusing. Tutton et al. (2004) raised an important issue in the context of UK Biobank's open consent procedures. Open consent was used because organizers could not determine all conceivable future use of the biological samples, but did state further consent would be sought for research outside the originally covered biological assays. The authors write: "It is a confusing mix of signals because, in practice, the stated purposes of UK biobank are so broad that it is unlikely that any research would be seen to be outside this original open consent" (Tutton et al., 2004: 284).

There has also been much work on the commercial and regulatory challenges of stem cell banking (Courtney et al. 2011), as well as the contradictory nature of value in the context of banking cord blood (Brown, 2013). The latter is explored by Brown through the conceptual lens of hope and promissory expectations. However, the specific role of patients, and the social and ethical issues around the storage and use of potentially sensitive health data from members of the public, is comprehensively reviewed by Haddow at al. (2007) in the context of a major national project called "Generation Scotland." This is a familial-based genetic study, based on the collection of ethically consented samples and data (Smith et al., 2006).[4] Generation Scotland involves the gene identification of complex diseases, such as cancer, heart disease, and mental health, through the recruitment of 50 000 family individuals aged between 35 and 65 years. Cancer, heart disease, stroke, mental health problems, as with many other common conditions, tend to cluster in families, so this provided the underlying rationale for the project.

The key objectives of the project were to: (1) recruit and phenotype a family-based cohort and identify genetic variants relevant to various complex diseases affecting the Scottish population; (2) engage and consult the public on social, ethical, and legal implications; (3) build research capacity and collaborate widely to share knowledge and best practice; (4) create a national health informatics platform to link study data with real-time health-care data, building on Scotland's fully electronic patient records; and (5) integrate with the clinical genetics community within the health service and conduct exemplar studies with identified protocols (Smith et al., 2006).

To take part in the study, an individual must have one sibling, preferably more, plus additional family members willing to participate. The participants (or probands as they are referred) are initially approached by their General Practitioner (GP) and then these individuals approach other family members. The study was launched in February 2006 and had recruited approximately 4,000 individuals by 2010. Individuals who take part in the study are asked to attend a clinical appointment for approximately two hours, where they donate a small amount of blood from which DNA is extracted and undergo a general medical examination. General information is also collected on their health and lifestyle, so this can eventually be linked to the genetic data. Participants are therefore asked to consent to their medical records (both past and future) being viewed by researchers. This is where the real value of the project lies. From building large-scale DNA databanks, and linking this to lifestyle and health data, it is suggested that significant benefits will emerge from increased knowledge to facilitate disease identification, diagnosis, treatment, and management of individuals at risk. In the long term, the hope is that drugs may be developed that will work in tune with an individual's genetic makeup and lead to less adverse drug reactions, so this is very much in the spirit of stratified medicine.

However, examples such as this do raise concerns about consent, privacy, and use of data. Haddow et al. (2007), focusing on the commercialization issues of these DNA databases, explored how this impacts on the willingness of patients and publics to participate. Drawing on the concept of "benefit sharing," and problematizing conventional approaches to notions of "gift" and "gifting," the authors highlight the social and cultural variations in understanding and expectations around benefit expressed by potential participants. In order to ensure broad health benefits from biobanks and DNA databases, commercial involvement is at some point inevitable. Yet, the underlying philosophy of donation, participation, and consent in the construction of these biobanks is often rooted in notions of altruism and gifting, which is in constant tension and flux with commercial norms. If participants feel their donation is a gift, which is provided altruistically, they may be less comfortable with the notion of commodification and profit in further downstream development of new therapies. Furthermore, many of these initiatives are based on principles of open access and innovation, as I discussed in chapter 4, which may conflict with pharmaceutical companies' desire for more exclusive data access to develop commercially viable products. Huzair and Papaioannou (2012), talking about UK Biobank, argue that knowledge from biobanks may in fact be subject to the tragedy of the commons and anti-commons, if patenting and scarcity turn it into a private rather than a public good. Haddow et al. focus on this prospect of future commercial involvement and how it can impact on individual's willingness to participate. Their findings, from ten focus groups on Generation Scotland, revealed that there was significant concern about future profit to pharmaceutical companies resulting from altruistic donation, regardless of the putative health benefits. However, the authors argued that "benefit sharing" or "profit pay off" models could be used to help accommodate both commercial and public interests and recognize their different values. This could help make conflicting notions of value and valuation more commensurable, in a way I have been suggesting throughout this book.

> The essential concerns revealed by our research distil around the injustice of using a gift to make a profit, the lack of institutional control over that profit, and the perceived disrespect to people by the commercialisation process itself. We suggest that many of these concerns may be traced to the current institutional set-up governing population genetic databases and other medical research. (Haddow et al., 2007: 28)

The authors argue that the institutional setup is often characterized by participants being relatively passive in the research enterprise, having little control over how donated material or data is used, and doing so on a putative "gift" basis, which suggests they have no expectation of future personal benefit. Furthermore, the researchers have few obligations beyond preventing harm to participants and there is no governance process to explicitly address concerns about commercialization (Haddow et al, 2007: 281). In order for these kinds of initiatives to have long-term success, and recruit willing participants on an ethical basis, a model of benefit sharing, which explicitly recognizes the importance of commercial involvement in therapeutic innovation, may be needed. Caulfield et al. (2014) suggest that the public is generally supportive of biobank research and recognizes its importance "...though the terms on which people will participate and the expectations they have regarding participation are not uniform and differ on key matters like benefit sharing" (Caulfield et al., 2014: 99). In particular, the general public may be less willing to participate, because commercial organizations may profit from their donation, than patients and their representative organizations, which have now a long history of collaborating with industry.

So, there is a major challenge in promoting participation of patients and publics in biobanks and genetic studies requiring the collection of personal data, and the management of expectations of future benefit is crucial to future success and the realization of long-term value. However, the future payoffs from biobanks are still largely ambiguous and uncertain. The sustainability of biobanks depends on them having long-term strategies and a secure funding steam. Ideally, they should also be able to link with other national or international biobanks. This requires interoperability through shared standards and processes for collecting, archiving, and retrieving data and samples. Tupasela and Stephens (2013) suggest that the fact that many biobanks will inevitably fail, be significantly restructured, or lose funding altogether, raises serious ethical concerns relevant to participation and its initial framing. This is captured in Henderson et al.'s (2013) national review and characterization of biobanks in the United States. The authors identified much diversity in terms of organizational structure, focus, and processes. Most of the banks were focused on cancer, followed by neurological diseases and HIV. Only 5 percent were for-profit, with the remaining either entirely publicly funded or "incorporated" firms. The authors argue that the increase in the number and diversity of biobanks, particularly since the emergence of genomics, raises questions about how best to classify them in such a way as to capture the subtle differences in policy, practices, and outcomes. Again, the lack of defined outcomes and shared practices can undermine long-term sustainability and benefits.

Cadigan et al. (2013), again looking specifically at the US context, argue that there is often a mismatch between the expectations of the managers of biobanks about how long the bank will exist and the time period for which funding is actually secured. Furthermore, many biobanks do not have a clear plan for what will happen to data and samples if the biobank closes and, crucially, many biobanks' resources are often underutilized by researchers:

> A recurrent theme in our case study interviews was concern regarding funding, specifically that funding would cease entirely or that it would be insufficient to cover costs. Despite this, most interview respondents said their biobank had no defined endpoint. (Cadigan et al., 2013: 4)

These discrepancies can serve to undermine public trust in biobanks, and may also work against the realization of long-term value, which itself is often used to build future expectations and increase participation.

In this section, I have unpacked just a few of the issues related to public participation in biobanks and data-linkage studies, which is crucial for the future realization of both patient value and economic returns from therapies based on new biology. Indeed, my interviews with pharmaceutical companies, public sector scientists, and policy-makers all revealed that translational biomedical research would be impossible without public and patient participation, often without any direct reciprocal benefit. This is important, as it makes patients and publics relevant to all stages of therapeutic R&D. I now, in the final section of this chapter, extend the discussion to the more recent emphasis on big health data initiatives, and the notions of value, waste, and efficiency that underpin them. This takes us beyond simply the collection and storage of data to how it may be used effectively to contribute to therapeutic innovation and patient benefit.

Expectations of Future Value from "Big Data"

In a 2013 report for McKinsey & Company, Kayyali et al. (2013) talk about a big data revolution that will transform US health care by accelerating the production of value and innovation. According to the authors, what is driving this interest are serious fiscal concerns about current health-care delivery. In the United States, health-care expenses are now over 17 percent of GDP:

> To discourage overutilization, many payers have shifted from fee-for-service compensation, which rewards physicians for treatment volume, to risk-sharing arrangements that prioritize outcomes. Under the new schemes, when treatments deliver the desired results, provider compensation may be less than before. Payers are also entering similar arrangements with pharmaceutical companies and basing reimbursement on a drug's ability to improve patient health. In this new environment, health-care stakeholders have greater incentives to compile and exchange information. (Kayyali, et al., 2013: 1)

This approach is also consistent with moves toward VBP, which I discussed in the previous chapter, where vast amounts of data may be needed to establish the true cost-effectiveness of new and existing therapies and provide evidential support for reimbursement decisions. I would also add that new biology, and the data associated with the novel therapeutic paradigms emerging from this research, disrupts conventional approaches to health care and the criteria for determining value.

The parallel emergence and priority given to evidence-based medicine should be seen in the light of this growing call to exploit big health data and implement the necessary governance systems and digital architecture. Together, evidence-based medicine and big data may provide radically new value frameworks for health care. However, value from big data will only materialize if there is an explicit intent to improve patient outcomes, rather than simply reduce health-care costs (Kayyali et al., 2013). So building the evidence base to support both improved diagnosis and treatment would seem to align well with this emerging emphasis around big data. However, evidence-based medicine itself has faced its own particular challenges in recent years. Greenhalgh et al. (2014) go as far as to describe it as a movement in crisis. When it first began to gain traction in the 1990s, there was concern that a focus on experimental evidence as a basis for clinical decision making would devalue basic science and the clinician's tacit knowledge, expertise, and experience, perhaps even undermining the close relationship between doctors and their patients. There was also skepticism that meta-analysis and systematic reviews of numerous, average clinical studies, which forms the basis for evidence-based medicine, could be used effectively to inform best treatment for individual patients who do not fit the average profile. Nevertheless, these concerns have subsided over time as many successful examples of evidence-based approaches have emerged. A booklet published by the organization "Sense about Science" listed 15 examples of successful evidence-based medicine (Sense about Science, 2013), and many successful studies are also cited in a report published by the Association of the British Pharmaceutical Industry (ABPI, 2013).

The notion that the accumulation and use of a broad range of evidence can inform decision making about optimal treatment for individual patients, which is the core philosophy of evidence-based medicine, is becoming largely accepted. However, there are persistent challenges. Greenhalgh et al. (2014) highlight the variations in how evidence-based practice is implemented and elucidate the many problems that continue to beset the movement. Three of these challenges are particularly relevant to the current discussion. First, the authors highlight the problem of there being too much evidence available. The number of clinical guidelines are simply becoming unmanageable as the range and scope of data increases. Second, there is an increasing emphasis on marginal gains determined from what the authors consider to be "overpowered," large clinical trials that tend to overestimate clinical benefit based on statistical significance. The authors argue that statistical significance in a large clinical trial does not always translate into significant clinical benefit for individual patients. Third, algorithmic rules tend to be overvalued in evidence-based medicine:

> Well intentioned efforts to automate use of evidence through computerized decision support systems, structured templates, and point of care prompts can crowd out the local, individualized, and patient initiated elements of the clinical consultation. (Greenhalgh et al., 2014: 2)

From this perspective, there is a risk that the individual patient experience and perspective becomes lost in the mass of aggregated data from multiple studies and PROs. In conclusion, the authors do emphasize the value of accumulated knowledge and systematic evidence from multiple empirical studies, but caution against steering evidence-based medicine too far towards population level statistics, risk-based analyses, and certainty claims that may be speculative at best. Webster (2007) also talks about how the "doing" of diagnosis can all too easily become "...lost in the codified and artificial world of evidence-based medicine" (Webster, 2007: 167). Evidence must always be capable of being individualized, and data should not usurp the importance and value of the clinican-patient relationship. Feldman et al. (2012) suggest that patient-centered care can emerge from big data, so long as the veracity of the data (its relevance, predictive value, and truthfulness) is acknowledged and firmly established. However, the authors argue that this has not so far been the central focus, as efforts have tended to be directed at establishing basic infrastructure to collect and collate data, and set standards for data sharing.

The question is how could big data contribute to delivering better health outcomes for patients and value for health systems, and what are the main challenges? The issues I have raised about evidence-based medicine do have implications for the future of big health data. How can we exploit the opportunities presented by inferential big health data to ensure evidence-based medicine supports innovative therapeutic development, while also making sure guidelines, treatment options, and outcomes meet the needs of individual patients and their expectations of value? Big data in the health bioeconomy is relevant at two different levels. On the one hand, it has been used in the context of collecting genomic and related data that may be used for future therapeutic intervention. Here, the key translational challenge is the development of tools powerful enough to analyze the vast amounts of data from both dry and traditional wet-lab studies. The harvesting and managing of this kind of big data is particularly important for studies in oncology, where public-private collaborations have emerged to create the infrastructures to collate, archive, and retrieve the data. This is very much about capitalizing on the data generated from biobanks and genetic linkage studies. Rubin (2015) points out that genomic data, and the technological barriers and complexities in linking this effectively to electronic patient record, means that few genomic reports can be found in electronic form. This is one major challenge for big data initiatives. On the other hand, big data relates to identifying optimal patient treatment, and value of therapeutic options, by building up the evidence-base (clinical and nonclinical studies) and aggregating vast amounts of real-time data, such that emerging trends become visible. This use of big health data is very much at the patient end of the translational bench-to-bedside continuum.

In its *Big Data Roadmap* report, the Association of the British Pharmaceutical Industry (ABPI) talks positively about the current and future role for big data in health care. It states: "Big data technologies make it easier to work with large data sets, link different datasets, detect patterns in real time, predict outcomes, undertake dynamic risk scoring and test hypotheses" (ABPI, 2013: 7). This approach could contribute to solving many of the challenges currently facing health-care providers, the bioscience research community, regulators, and industry, as I have discussed throughout this book. The report outlines the different types of data that provide opportunities for improving health care. These include: (1) health-care data from electronic health records and biometric sources; (2) genomic data, especially gene sequence data; (3) automated sensors and smart devices; (4) self-generated data and digital engagement, including PROs and online health forums; and (5) public data releases, such as academic, commercial, and governmental data that are made public (ABPI, 2013: 11). Together, all of these data sources are capable of generating big data that could be harnessed for broad public health benefit. In the United States, both the FDA and the National Institutes for Health (NIH) have set up big data initiatives. The FDA established three major initiatives in 2013, including "Open FDA" to help develop data mining techniques and use of big data to monitor safety and efficacy of new drugs. The NIH launched the Big Data to Knowledge (BD2K) initiative in 2012 to better harvest and exploit biomedical big data.[5] The BD2K initiative has four key aims, which, in combination, the NIH expects will enhance the utility value of biomedical big data. These include facilitating the use of biomedical digital assets by making them more accessible, conducting research to develop new methods and tools for big data analysis, enhancing training in big data science, and supporting an evolving "data ecosystem" to accelerate drug discovery.

The ABPI, and other supporters of big health data like the FDA and NIH, sees opportunities and value across the whole life sciences and health-care R&D pathway, from discovery research in genomics all the way to health outcomes in the clinic. One key benefit highlighted in the ABPI roadmap is the ability to exploit greater efficiency in health care and reduce waste. So this brings us back to the broader notion of value and waste in the health innovation ecosystem, as I described in chapter 1 and which I critically explore in the remainder of this chapter.

The notions of value, waste, and efficiency are every present in discussions about the evolving health innovation ecosystem and the promise of big data. Chalmers et al. (2014) talk about inefficiency and waste in terms of setting research priorities and identifying research that has already been conducted. Recognizing that some "waste" is unavoidable, the authors believe there is always avoidable waste that could lead to greater value and efficiency in health. They point to the underutilization of existing research, often caused by a failure to identify research that has already been done:

> Systematic assessment of what is already known or being researched is essential when decisions are made about what further research to do. Such assessment will identify what should be replicated, avoid

unnecessary duplication, and result in research that addresses defi-
ciencies in previous work. Although the point at which necessary rep-
lication becomes wasteful duplication can almost always be disputed,
decisions should be informed by as high a proportion as possible of the
relevant existing evidence. (Chalmers et al. 2014: 159)

Here, big data could be used to better identify and utilize existing
research to ensure no publicly funded research is wasted by being
underutilized as a resource to inform future work. This is part of
the NIH and FDA initiatives. However, as Glasziou et al. (2014)
argue, waste can also emerge from incomplete or unusable reports
of biomedical research, so even if relevant studies are identified and
reviewed as part of a big data initiative, there may be little value if the
data lacks interoperability. Chan et al. (2014) also point out that a
great deal of research is inaccessible because full data sets may not be
published, in part or in whole. This is a major challenge for those who
wish to exploit the tools of big data to address health issues, because
full data sets must be accessible for any value to be identified and
exploited. The authors point out that half of health-related studies are
unpublished or unreported, and very few study protocols and participant level data sets are available. Furthermore, if we take this beyond
publicly funded research, and include commercial research, the challenges are perhaps even more significant. Most commercial research
is not published, and there is high variation in the tools and standards
used in studies and database architecture. This makes it difficult to
integrate data from many organizations. Also, most pharmaceutical
companies do not generally share data on failed drug programs, even
though this could be highly valuable in a big data context. It could
be used to both improve public health and safety and reduce overall
costs of drug discovery.

   What this tells us is that big data could, potentially, be used to
generate value for therapeutic innovation and improve patient outcomes, but the challenge is generating data of sufficient value in the
first place, and ensuring it is accessible and transferable. Also, big
data might in theory provide an opportunity to add value to existing
research and create efficiency, but if the data is not sufficiently interop-
erable and capable of directly informing treatment protocols, it could
generate inefficiency and be potentially wasteful. Furthermore, a failure to truly harness the value of big data would be to underutilize
the numerous studies that have required voluntary participation by
patients and publics. The fact that big data are being more widely discussed in the context of value, waste, and efficiency is also interesting

in the context of the health innovation ecosystem. How useful is the concept of "waste" in the context of health innovation? In chapters 1 and 4, I challenged the notion that translational experiments in organizational restructuring of R&D should be considered to have failed if they do not meet the promissory expectations generated at their outset. I argued that waste is an inevitable part of any innovation ecosystem, and this is part of the trial-and-error process that is a defining feature of experimentation. Also, if we embrace the broader concept of value that I have been arguing for throughout this book, the role of big data and the very concepts of waste and efficiency may need to be rethought. Big data does have the potential to deliver a wide range of benefits, and value, to industry, health-care systems, and patients. However, the realization of that value will depend on the existence of appropriate systems of regulation, governance, and public and commercial innovation support systems capable of driving it forward. In this context, the role of patients and publics, as both sources of the tissue samples and data that is needed to build a big data strategy, and users/beneficiaries of health-care services, will be critical to success.

Conclusion

In this chapter, I have provided a critical reflection on the changing role of patients and publics in the contemporary health innovation ecosystem, building on previous chapters that focused on innovators (both commercial and public sector) regulators, and policymakers. It is clear that the role of the patient has undergone profound change over the past three decades. New biology, and particularly the mapping of the human genome, has shifted the patient from the margins to the center of innovation, from passive recipient of health care to active participant in R&D.

Patients and publics are now a major social, political, and economic force in health innovation, not only at the individual level of participation in experimental clinical trials, but also through collective lobbying through patient organizations. As such, they have helped shape the broader health ecosystem. I have tried to highlight in this chapter the different levels at which this participatory agenda has been expanded, and drawn attention to some of the key challenges and opportunities for current and future value realization. The emergence of big data initiatives is the latest enactment of this reimagining of the patient and public role. As a means to enrich data collection and analyses through which decision making on treatment efficacy and value can be improved, big data is clearly high on the agenda of both policymakers and industry, and will likely define the continued evolution of the life sciences and new biology in the next few decades. However, the future trajectory of life science innovation, and its impact on meeting the health-care needs of very changeable national and international health-care expectations, will depend on how these new big data initiatives, and broader public and patient involvement in research, align with changing opportunities and challenges in other parts of the health innovation ecosystem.

Chapter 7

Rethinking Value and Expectations
in the Health Bioeconomy

Introduction

In this book, I have proposed a new approach to exploring the enactment of value in the context of an emergent health bioeconomy.
Specifically, I have attempted to provide a broad and systemic analysis of the constitutive elements and key drivers of what is a complex
health innovation ecosystem, with multiple stakeholders each operating with different notional ideas of the benefits, limitations, and long-
term value (beyond merely the economic) of various Research and
Development configurations and options. I have also tried to think
through how different expectations are managed, and new organizational routines and practices made to work, as "new biology" promises to deliver radically new therapeutic paradigms, but is confronted
with a confluence of technological, regulatory, and social challenges.
By tracing the evolution of new biology and the emergence of what
we might call a health bioeconomy in the twenty-first century, I have
been able to investigate and critically reflect on how the institutional
and organizational landscape for health R&D, particularly within
Europe and the United States, has undergone a profound transformation as a result of the emergence of new science and technology
options, and multiple types of value and valuation practices.

A new regime for biomedical innovation has emerged in the past
two decades, and this must be seen in the broader light of a perceived
set of problems that have beset contemporary health innovation—
in particular, the parlous state of the large pharmaceutical industry
over the past two decades due to a perceived "productivity crisis,"
and related concerns about the sustainability of the blockbuster drug
model and the economic value assumptions that have underpinned

it. This has led to the emergence of many promissory expectations of future value built on a powerful trope of "translation," which has been enacted by both commercial and public sector organizations and has had a material effect on conventional R&D practices. It has also disrupted traditional and entrenched boundaries between the disciplines, professions, and industrial sectors that comprise the complex health innovation ecosystem. A particularly salient phenomenon has been the increasing role of public sector actors in driving innovation and attempting to realize different types of value.

In this final concluding chapter, I first provide a brief synopsis of the key arguments presented in each chapter and suggest how, together, they help throw light on the nature of the fragile health innovation ecosystem and the mobilization of value, and valuation practices, within it. I then conclude by thinking through what the empirical data I have presented, coupled with the conceptual approach adopted, tells us about the future evolution of the bioeconomy and therapeutic product development. The key question to be addressed in this chapter is: *What is the future for therapy in light of the many experiments in translational medicine, the nature of the evolving bioeconomy, and the constellation of value therein?*

A Summary of the Argument

In chapter 1, I outlined the general conceptual framework for the book and its key empirical foci. I began by illustrating what is novel about new biology and the bioeconomy, and why I believe it compels us to rethink our conventional, and highly circumscribed, notions of value. The question I raised was: *What, if anything, is distinctive about "new biology" and the "health bioeconomy" and how do they challenge conventional systems of health innovation and the enactment of values?* In terms of "new biology," I argued that its distinctiveness lay in the interdisciplinary nature of its knowledge base and highly distributed organizational structure, in which the public sector has a pivotal role beyond merely financing science or creating incentives for innovation. Collaboration has been a defining feature of new biology to date, and played a major role in driving its progress in particular directions. For these reasons, its intellectual and industrial development, and durability, has been very different from, say, the physical sciences, which dominated science policy and societal discourse for most of the twentieth century. The concept of innovation ecosystem enables us to better unpack and analyze the systemic characteristics of health innovation based on this new biology. It also helps us to better understand the nature of the emerging and still quite fragile health bioeconomy, which underpins research and industrial policy.

This bioeconomy, which has been widely critiqued within social science because of its putative neoliberal underpinnings (Birch and Tyfield, 2006), is also supported by a number of policy assumptions that have driven particular kinds of innovation options and expectations of future value (Borup et al., 2006). While accepting the gap between speculative and actual economic value—and the hopes, expectations, and performative function of neoliberal policies that have driven some aspects of the bioeconomy (Hilgartner, 2007)—I argued that the constitution of the bioeconomy in the twenty-first century opens up the possibility of studying value in a much broader, and perhaps more useful sense. Drawing on recent work on valuation as a social practice (Helgesson and Muniesa, 2013), and those authors who have encouraged us to recombine, or at least problematize, the conventional distinction between economic and noneconomic values (Fourcade, 2011; Heuts and Mol, 2013; Stark, 2009), I outlined a broad conceptual approach to value that could more meaningfully unpack the foundations of the health bioeconomy and help us better understand the evolution of the health innovation ecosystem. In summary, I wanted to extend the definition of value, and the different valuation practices deployed by various actors and institutions, beyond narrow econometric principles and matters of finance. Many critiques of the bioeconomy are based on the notion that it simply commodifies life, and monetizes certain relations and practices around health such that a broader range of values are often marginalized or entirely neglected. However, the bioeconomy, and the health innovation ecosystem, is suffused with multiple notions of value and valuation practices, not all of which are wholly or predominantly economic in nature. Yet they are all crucial to the maintenance and future success of the innovation ecosystem. Since the bioeconomy encompasses many public and commercial organizations and actors, we need a vocabulary and conceptual framework that captures the multiplicity of value constituted within this type of innovation ecosystem. This framework must be capable of capturing the diverse, and sometimes conflicting, social, institutional, and organizational relations which, together, generate both tangible and intangible benefits from health-related R&D. It was from this starting point that I was then able to begin outlining the various elements of the health innovation ecosystem and tell the story, using illustrative case examples, of how new biology has both challenged conventional approaches to therapeutic development and created new opportunities, and challenges, for this broader realization of value.

In chapter 2, I began that story by reflecting on the history of pharmaceutical innovation, and the industry dynamics that have accompanied it, to contextualize the distinctiveness of new biology and its impact on what was an indomitable pharmaceutical industry at the end of the twentieth century and early twenty-first century. This chapter addressed the question: *What impact have the life sciences had on the organizational structure, commercial strategies, and R&D practices of the pharmaceutical industry?* In this chapter, I highlighted how interdisciplinary new biology, and specific life science technologies, presented both opportunities and challenges for an industry that had built up capabilities over many decades in small-molecule drug development. As these conventional "blockbuster drugs" reached maturity in the mid-1990s (Mittra et al., 2011), and R&D pipelines began to dry up, the promise of future economic value based on a disruptive life science trajectory was an attractive proposition to the multinational pharmaceutical industry. All the major companies began investing in new screening technologies, reorganizing their R&D processes and management structures, and exploiting mergers, acquisitions, and strategic alliances with smaller biotechnology companies to capture new knowledge and expertise. R&D was transformed from a largely craft-based process to a more robotic approach that tried to exploit new economies of scale (Nightingale, 2000). These experiments in R&D management were driven by both anxiety about the parlous state of drug pipelines (the productivity crisis), and high expectations of a new paradigm for therapeutic innovation that, it was hoped, would generate long-term value.

However, new biology was, and continues to be, a double-edged sword for the pharmaceutical industry. On the one hand, it complemented conventional drug discovery (through better screening capabilities), but on the other hand it was seriously disruptive to the traditional blockbuster model. Therapies based on new biology require very different business models, and have uncertain and risky routes to market, which necessitate the creation of new value chains and regulatory systems to enable them to successfully transition from bench to bedside. Furthermore, valuable knowledge and capabilities in life sciences are highly distributed, so no one company can do everything "in-house." In this context, new biology presented the pharmaceutical industry with numerous options for value realization, but also major challenges in exploiting these alongside conventional drug discovery and development. Indeed, much of the initial hype, which suggested that genomic technologies would transform the fortunes of the pharmaceutical industry, turned out to be wrong or at least premature. However, as firms struggled to adapt to the life sciences, and increasingly sourced knowledge, technology, and expertise from smaller companies in the 1990s, the very nature of the innovation ecosystem underwent a profound change. In particular, the diversity of actors and organizations needed to successfully do life science R&D, and the emergence of new value drivers, enablers and constraints, rendered this a particularly important juncture in the contemporary history of therapeutic innovation.

This snapshot of how the pharmaceutical industry responded to the opportunities and challenges of new biology at the turn of the twenty-first century provided the background context for chapter 3, where I explored the "broken middle" of health innovation and the emergence of "Translational Medicine" (TM) as a driver of much broader change in R&D processes and practices within the health bioeconomy. The question I addressed in this chapter was: *What perceived challenges, opportunities, and practitioner values in health innovation have driven a new translational policy agenda, and with what consequences for the bioeconomy?* This chapter took the story beyond the narrow interests of the multinational pharmaceutical industry, and their concerns about R&D efficiency and maintaining shareholder value through "blockbuster" drugs. Here, I considered in more depth how the policy community and various noncommercial actors and organizations have, through an emerging translational agenda, become an integral part of the health innovation ecosystem. In so doing, they have driven particular enactments of value, shifted the debate as to what is valued in health and research, and become enrolled in a diverse set of promises and expectations about the future of health innovation and the bioeconomy. This has all taken place in the context of a current system that is perceived to be "broken," captured through the populist rhetoric of the "valley of death" (Nature, 2008). Indeed, this has become a powerful trope for describing the current malaise in drug innovation, as well as providing justification for the series of experiments in organizational restructuring that have ensued.

In this chapter, I critiqued the crude, linear model of innovation, and the conventional distinctions between basic and applied research, before looking at organizations like the National Institutes of Health (NIH) as anticipatory and promissory organizations for TM. They are emblematic public sector organizations that are transforming what it means to "do R&D" in the twenty-first century. I highlighted how many of the practices, and R&D challenges, underpinning the broad TM agenda, which itself is diverse and often contested by practitioners, are not entirely new. However, I also suggested that TM has materially affected the structure and organization of the health innovation ecosystem and the configuration of value in the broad bioeconomy. It has also led to certain research areas, such as the identification and validation of novel biomarkers, becoming a priority for research funding and the building of new collaborative research consortia. Interviews with key practitioners also illustrated the many definitions and framings of TM, and the notional ideas of value and worth that are ascribed to these ongoing endeavors.

The more microlevel impact of this TM agenda was explored in much more detail in chapter 4. Here, I addressed the question: *In what ways has the "doing" of R&D been reshaped by the institutional and organizational restructuring precipitated by translational policies and how are stakeholder expectations and values recognized and managed?* I argued that interdisciplinary knowledge and research, which is being prioritized within these new collaborative partnerships for R&D, has disrupted conventional professional and disciplinary boundaries. Drawing on case examples of three public-private partnerships, I revealed how different ideas of value and benefit, and long-term expectations, are managed as actors constituted within these collaborative models "muddle through" to make R&D work in practice. Actors working within new organizational regimes negotiate, trade, and attempt to realize a diverse range of values and benefits, only some of which are economic in nature. The new health bioeconomy, if it is to be successful, must enable these different forms of value to accrue to different actors and stakeholders in the broader innovation ecosystem. It must also take account of the fact that actors and organizations may be subject to a range of valuation cultures and standards. The same practice, and potential benefit, may be valued very differently depending on where one is institutionally and culturally embedded in the innovation ecosystem. For instance, what a large pharmaceutical firm ultimately values in the context of a collaborative research project may differ from an academic scientist or a health-care provider. Yet, these different valuation cultures, and the metrics for determining success, must be made commensurable for R&D and innovation to work. The twenty-first-century health innovation ecosystem must continually manage the clash in cultures and norms between professions, disciplines, and institutional actors to make R&D durable. Pisano (2006) had highlighted the tensions between the worlds of science and business in his highly cited book,

Science Business, but I extended this to a broader range of actors, professions, disciplines, institutions, and organizations embedded in the modern health innovation ecosystem.

In chapter 5, my analysis moved beyond the core innovation communities (those producing new knowledge, technology, and products for the bioeconomy) to consider the impact of regulation and policy on health innovation. Here, I revealed how new biology challenges not only innovators, but also the often sclerotic regulatory and reimbursement systems through which new therapies must pass in order to reach the market. The question addressed in this chapter was: *How has new biology both challenged and transformed conventional regulatory systems and the resilience and adaptive capabilities of health-care systems to innovative therapies?* Using the case studies of regenerative medicine and stratified medicine, I demonstrated how these disruptive and potentially path-breaking approaches to therapy, or what some now usefully refer to as novel "bio objects" (Holmberg et al., 2011), challenge regulatory systems and health-care pathways, which have built up incrementally over many decades to accommodate conventional types of therapy, namely small-molecule drugs. Although regulatory agencies in both the United States and Europe have attempted to adapt particular elements of the regulatory system to try to accommodate path-breaking therapies (EMA, 2011; FDA, 2011), and think through new approaches to regulatory science, I argued that this is often piecemeal. In order to generate the promissory value and full patient benefit of new, advanced therapies, regulatory systems may need to undergo a much more radical change and become less precautionary in their approach.

The so-called golden age of drug discovery from the 1940s to 1960s was conducted with minimal levels of regulatory oversight. Although stricter regulation was necessary in the wake of the thalidomide scandal, we have witnessed a continual ratcheting up of regulatory standards, and sometimes the emergence of de novo regulatory protocols for specific kinds of technology, to the point that it negatively affects innovation. This has a particular impact on smaller, innovative companies that are at the forefront of developing applications based on new biology (Tait, 2007). At the end of this chapter, I addressed the fourth hurdle of health technology assessment and reimbursement as a major barrier to the development of regenerative and stratified medicine. I also considered whether value-based pricing (VBP) might enable the full potential of these new approaches to therapy to be realized. VBP, I suggested, defines value in the much broader sense I have been advocating for throughout the book. If it were to be fully implemented, it could potentially enable radically new therapies, which may bring major patient and social benefits, to breakthrough into the clinic. At the moment, value and reimbursement is based on crude and narrow economic metrics, and quality-of-life criteria that seem inappropriate for many of the innovative therapies being developed. This approach, often referred to as cost-effectiveness, is highly restrictive and may be preventing the broad societal benefits from new biology being fully realized. However, it appears that a more radical form of VBP is unlikely to be implemented, and indeed VBP itself is a contested term that can mean different things to different people. Because value is such a subjective concept, which emerges as a product of specific social and institutional valuation practices, VBP could be used instrumentally to either increase or reduce the price of therapy. Its potential benefits to patients, industry, and health-care payers will therefore be highly varied and could perhaps be a source of continual friction.

In Chapter 6, I built on this discussion about value and benefit of new therapies to think about the role of patients and publics in contemporary health innovation. Patients and the general public are key to all stages of health innovation and have been transformed from passive recipients of health care to active participants in the R&D process. The question I addressed in this chapter was: *what are the implications of the changing role of patients and publics in the new health bioeconomy, and how can their expectations and values be better understood and managed?* Drawing on a broad literature that has highlighted the value of patient's experiential knowledge (Canon-Flinterman, 2005; Pols, 2014) and the emergence of a patient-centered approach and political advocacy (Epstein, 1996 and 2007), I began by showing how patients have become a major political and economic force in healthcare policy and therapeutic innovation. They now contribute more directly to the medical knowledge base and can have an instrumental impact on what kinds of treatments are developed and eventually used within the clinic. The increasing use of patient-reported outcomes in clinical research is illustrative of this growing influence of the patient as a key actor in the innovation ecosystem, who may be valued in a number of different ways for a number of different purposes. Of course, there are challenges to incorporating the so-called patient perspective, and it was important to critically reflect on how patient participation might improve medical knowledge and health care without treating individuals instrumentally as a means to an end.

I then explored the role of biobanks, and the related emergence of big health data, as a future driver of the health bioeconomy, which is generative of a diverse range of promissory expectations. In particular, I reflected on how the emergence of these data repositories and their complex governance frameworks structure public participation processes and try to take account of their expectations around consent and benefit. To summarize, the future of the health innovation ecosystem and the bioeconomy will depend on the continuing participation of patients and publics in health innovation processes, including their important role in donating biological material and data/ information, some of which may be highly personal. In order to guarantee their participation and consent, the value and promised future benefits must extend beyond narrow economic and commercial criteria. Value must therefore encompass broader clinical, scientific, and social criteria, in addition to the conventional economic bottom line that has tended to be prioritized in innovation policy, and which has driven commercial drug development for many decades.

Rethinking Value and the Innovation Ecosystem

Each of the individual chapters of this book, taken together, highlights the changing nature of value, and its underlying social relations and practices, within an ever broader and expansive health innovation ecosystem. They also reveal how different elements or sectors within the innovation ecosystem have evolved and been reciprocally shaped by the emergence of new biology and its associated technologies. Over the past three decades, we have witnessed a prolonged period of experimental policy initiatives, commercial strategies, and major organizational restructuring, as a direct result of new biology and its promise of numerous benefits and value to industry, medicine, and society. Enacted through the emergence of new "translational" policies, this has long-term consequences and implications for all three of these sectors, each of which is having to rethink what is of ultimate value in health innovation and therefore worthy of support.

First, the pharmaceutical industry has been compelled to explore alternatives to blockbuster drug development, and the economic norms and value propositions that drove this highly successful, but unsustainable, strategy from the 1950s to the 1990s. They have sought value from knowledge, technology, and expertise emerging from the life sciences, and the small companies and public sector research organizations that are pioneering new health-care innovations. Although the economic bottom line does, obviously, still drive the strategies of the commercial pharmaceutical industry, the involvement of this industry in numerous precompetitive collaborations and networks, and the use of new incentives that have been created for the industry to tackle areas of unmet medical need, suggests that it is beginning to entertain a broader notion of value. Some of this value includes elements that are not intrinsically or predominantly economic in nature. Even though commercial firms are ultimately driven and judged by shareholder value (which is fundamentally driven by economic returns on investment), this can only be realized by understanding, capturing, and appropriately managing a broader set of values and expectations that are constituted within the very broad health innovation ecosystem.

Second, both policymakers, which have largely driven the translational agenda, and the academic and clinical research community, have begun to embrace a broader range of values in order to operate effectively within the new organizational regimes that the health innovation ecosystem now demands. As public sector funding is increasingly earmarked for research that contributes to the translational agenda (i.e., promises a clear pathway to clinical or social impact), the definition of what is valued in academic research has also shifted significantly. While conventional, blue-skies academic research that can be published in top-tier academic journals is still the principal criteria for success for many academics, and the *lingua franca* of most university departments, there is now recognition that collaboration with industry in interdisciplinary, translational research also has significant value and is worth pursuing for a variety of reasons. Indeed, partners in collaborative research, which these days now often includes health service providers, will always have different expectations of what value and benefit may ultimately emerge from such collaboration. However, what I have shown in this book, through case examples, is that multiple benefits can be accommodated if expectations are realistic and managed effectively. Furthermore, success or failure of particular initiatives should not be judged on the basis of a narrow set of evaluative criteria, especially if these are predominantly economic in nature. The fact that the health bioeconomy is not perhaps living up to its financial expectations, which is often highlighted as part of a broader critique of its underlying activities, does not mean that it has no real value and is not worth nurturing. In this context, the Dutch valorization model, which I discussed in chapter 4, is particularly useful in teasing out these broader notions of value and benefit. Future success in health innovation (defined as bringing better therapies to patients and health systems) will not come from short-term strategies based on immediate economic returns. Indeed, for many innovative therapies, such as regenerative medicine, any major economic gains will likely materialize in the future and will be based on current collaborative activities that are driven by a much broader set of values and expectations. Of course, this uncertainty over future economic sustainability does impact on current commercial investment in these technologies, and is one reason why large pharmaceutical firms have shown limited interest in developing the technology. Nevertheless, precompetitive collaboration based largely on noneconomic criteria (at least in the short term), and the involvement of a range of public and private actors and sources of finance, does continue apace. These kinds of models are likely to continue as disruptive technologies struggle to navigate a clear route to the clinic and exploit existing value chains.

Third, the increasing recognition by regulators that they have a direct impact on the success or failure of particular innovation options suggests that they too are beginning to embrace a broader notion of value, beyond merely the value of safe and effective medicines. They must balance the requirement to protect public health with the need to ensure there is a viable industry to develop new therapies for patient benefit. Although, as I have argued, regulators have perhaps not been bold enough in taking a more proactionary stance to path-breaking technology, the emergence of such innovations as adaptive licensing and conditional approval suggests that they are moving in the right direction. The fact that organizations responsible for health technology assessment and reimbursement, which are part of the regulatory system in the sense they serve as gatekeepers to the clinic, are now at least talking seriously about VBP further supports my argument for a much broader conceptualization of value. Of course, as I discussed in chapter 5, the devil will always be in the detail. The extent to which policymakers and regulators will fully embrace a radical approach to value in the pricing and marketing of new therapies is still very much uncertain. Nevertheless, society will unlikely tolerate a continuation of the current pricing structure for pharmaceuticals, and its very narrow and instrumental approach to value.

Finally, the increasing role of patients and publics in health innovation, and their emergence as a powerful political and economic force, means that they have a significant influence on what is ultimately valued in the health bioeconomy. As they donate biological material, participate in clinical studies, and consent to their data being used to improve drug development and treatment regimes, patient's experiential knowledge and engagement with health innovation systems has acquired real value that could potentially be used to satisfy broader societal expectations. On the one hand, this may be read cynically as a further commodification of the patient, as evidenced in the way illness narratives have become tradable products in global health markets. But on the other hand, the routine use of patient reported outcomes suggests that the patient perspective also has major clinical and regulatory benefits. The role of politically active patient organizations, many of which directly fund research, also influences what can be achieved in health innovation. They can mobilize resources, knowledge, and people such that value in health innovation is recalibrated in line with the needs of patients and broader society, at least in theory, if not always in practice.

What was clear from my arguments in chapter 6 was that patients, and to an extent other publics, do have quite a sophisticated understanding of the health innovation ecosystem, and recognize that this encompasses a constellation of values and valuation practices, including economic ones. Although various accounts and enactments of value may at times be in conflict, and can be a source of tension between different actors and stakeholders, they are not mutually exclusive. Patient groups and health-care providers, in particular, recognize that multiple expectations of future value and benefit must be met if progress is to be made in tackling the major health innovation challenges continuing to face society.

The Future of the Health Bioeconomy

So, given the arguments presented in this book, and the approach to value I have been arguing for, what is the future for the health bioeconomy and therapeutic innovation? Societies continue to face a number of grand challenges, and science and technology is now routinely exhibited as an important part of any long-term, sustainable solution. In the context of health, we have aging populations, significant areas of unmet medical need, and a pharmaceutical industry still in the throes of a productivity crisis. Recent concerns about antimicrobial resistance, and how to deal with this at both a national and global level, are just one example of the many societal challenges ahead. The traditional blockbuster business model is clearly coming to an end, but the radically new therapies promised after the successful mapping of the human genome have yet to materialize in any great number. Furthermore, there are major cost constraints on health service providers at a time when new, advanced therapies are incredibly expensive. So value is becoming an increasingly salient issue within health-care communities.

In this book, I have identified many of the reasons for the continuing gap between the promise and the reality of path-breaking advanced therapies, paying particular attention to the regulatory systems, and challenges of building new business models and pathways to clinics that do not yet have the required institutional readiness to accept therapies such as regenerative and stratified medicine. All of these challenges are significant, but not insurmountable. It would be surprising if new biology did not eventually deliver on most of the major promised benefits and value to patients and health systems. However, a greater understanding of the broad, systemic aspects of the whole innovation ecosystem, coupled with much greater recognition of value beyond the strictly economic, is needed to ensure a sustainable bioeconomy is nurtured and enabled to potentially deliver benefits to industry, medicine, and society at large.

At the beginning of this book, I stated that my aim was to think pragmatically about the key challenges facing the long-term realization of value in the health bioeconomy, and the sustainability of the health innovation ecosystem. I also sought to problematize both those accounts that dismiss new biology as mere hype, or consider its impact on R&D norms, practices, and organizational principles as marginal, or perhaps even negative. Instead, I have drawn on a range of insightful accounts and conceptual approaches that have shown how R&D policies and practices are shaped by promissory visions and expectations of future value and worth. In the context of translation, I have demonstrated that a perception of a broken health innovation system has led to the instantiation of a range of policy initiatives that have materially affected the practice of R&D, and the ways in which value is conceived by different organizations and actors. Critiques of the bioeconomy, and skepticism about the increasing role of public sector money and resources in commercial innovation processes, tend to invoke a narrow definition of value, rooted in corporate economic interests. In this context, new policy approaches to drive innovation and contribute to the "translational" challenges of drug development are often portrayed as part of a neoliberal, corporatization of health and policy. My approach, which expands the scope of value and benefit and problematizes the conventional boundaries between public and private, and corporate and non corporate entities, helps us better understand the myriad practices underpinning the modern health bioeconomy and may provide better justification for particular kinds of policy intervention. Mazzucato (2013) highlights the importance of what she calls the "entrepreneurial state," particularly in the context of meeting innovation challenges caused by market failure. The state, in these cases, must provide direct financial support and create incentives for commercial innovation where such incentives appear to be lacking. However, missing from her account is recognition of the largely noneconomic role public sector institutions play in modern health innovation, and the different types of value and valuation practices that determine the evolution of the health innovation ecosystem and its underlying bioeconomy. In this book, I have attempted to outline this much needed broader and systemic approach.

The many experiments in translational R&D, which are continuing apace, will have varying degrees of success, but in the context of an ecosystem, we should perhaps be careful to avoid labeling any of these experiments an abject failure. Future work in this area might continue to explore the evolution of the health innovation ecosystem through the broad conceptual lens of value and valuation practices. This helps us unpack and better understand how different actors and organizations work together to make R&D work in practice and, hopefully, deliver tangible benefits to patients and society, as well as commercial and noncommercial innovators. I have tried to begin developing a more sophisticated approach to value, and link it to empirical data in the context of health innovation systems. The take-home message is that health innovation ecosystems and the bioeconomy are suffused with many different types of value, and driven by multiple actors and organizations with very different expectations of future benefit from particular R&D configurations and practices. Only by recognizing and taking seriously these differences, and looking at how various groups muddle through to make R&D work in practice, can we fully understand how specific enactments of value create different kinds of benefit for the key stakeholders throughout the diverse health innovation ecosystem.

Notes

1 New Biology and the Foundations of a Health Bioeconomy

1. "New biology" is generally defined as the branch of biology that is focused on biological phenomena at the molecular level (through the study of DNA, RNA, proteins, and other macromolecules). "Life sciences" is conventionally understood to refer more broadly to any natural science in which the primary object of study is living organisms, so it is not restricted only to molecular biology. In this sense, new molecular biology is a subdiscipline within the life sciences. However, in this book I use both terms interchangeably, because key practitioners and stakeholders in the sector do not generally maintain strict definitional boundaries, and use both "new biology" and "life sciences" to describe experimental work in molecular biology and all related knowledge, techniques, and technologies being developed to industrialize and capture value from fundamental research on life. However, I elaborate a broader use of the term "new biology" in order to capture the distinctive organizational and institutional structure that envelops this particular type of interdisciplinary life sciences.

2. Recombinant DNA technologies allow for DNA from two or more sources to be removed (using restriction enzymes) and transferred into the plasmid of a bacterium to produce a recombinant protein molecule, which may be used as therapy.

3. A value chain is the set of activities that a company, or a set of companies and organizations, must perform in order to take a product from conception to final market and generate "value." In the contemporary life sciences, value chains tend to be distributed across many organizations, each with their own particular business models. The health innovation ecosystem encompasses these multiple value chains and business models.

4. Modern drug development is split into a number of development phases, which are often presented as a linear process from discovery of a new therapeutic compound (traditionally a small-molecule), through to identification and validation of the drug's biological target (an

enzyme or receptor, for instance), and then on to further preclinical development, which involves all the early animal and in vitro studies that must be conducted to establish proof of concept and early indications of safety. The preclinical phase takes approximately five years for completion. Then comes the three main clinical phases, which is the longest and most expensive part of drug development, often taking up to ten years for completion. Conventionally, phase 1 clinical studies are "first in human" trials to determine the metabolic and pharmacological actions of drugs in humans, the side effects of increasing dosage, and sometimes to gain early evidence of efficacy. They most often involve a small number of healthy volunteers (up to 50), but may also involve small numbers of patients in certain circumstances (e.g., for highly novel therapies or orphan disease areas). Phase 2 clinical trials are designed primarily to ascertain drug efficacy (how well the drug performs under highly controlled conditions), as well as continuing phase 1 safety studies. Phase 2 trials generally involve between up to 300 patients, but may be as low as 20 for some advanced therapies or orphan diseases. Phase 3 clinical trials are randomized-controlled, multicenter clinical studies to evaluate the drug's efficacy for a particular disease indication in a large sample of patients (sometimes as many as 3,000 or more for many small-molecule drugs). They often use one or more existing treatments as comparators. This stage also aims to identify potential side effects and risks associated with the treatment. Postmarketing surveillance, which is often referred to as phase 4, is then used to assess the drug's clinical effectiveness (i.e., how well does the drug perform in the clinic where conditions cannot be controlled as they can in a formal clinical trial). This is a very crude summary of the drug discovery process, which makes it appear to be relatively inflexible and linear, but the way drugs move through the system is changing, as I describe later in the book. Also, there are many parallel processes and feedback loops in drug development that are not captured in the simplistic linear model.

5. Conceptualizing life science innovation in terms of an ecosystem is useful in that it emphasizes the interdependencies between different actors, institutions, and firms operating in the spatially and temporally distributed innovation process and, crucially, captures the evolutionary-like mechanisms that determine success or failure of different product development strategies and firms within a sector. For an in-depth description of a unique approach my colleagues and I at the Innogen Institute have developed to study and analyze value systems in regenerative medicine, which is rooted in the notion of an innovation ecosystem; see Mastroeni, Michele, Mittra, James, and Tait, Joyce. (2012) *Methodology for the Analysis of Life Science Innovation Systems (ALSIS) and its Application to Three Case Studies*, TSB Regenerative Medicine Programme: Value Systems and Business Models REALISE project Final Report, May 29, 2012.

6. Although The HGP tends to be presented as a single unified project, there were competing groups attempting to map the human genome, and genome sequencing has a long history that is not so easily captured by the simple descriptor "The Human Genome Project." For a rich history of gene sequencing and the many diverse actors, disciplines, and organizations responsible for its development, see Miguel García-Sancho's 2012 book, *Biology, Computing, and the History of Molecular Sequencing: From Proteins to DNA, 1945–2000* (New York: Palgrave Macmillan).

2   Crisis in the Pharmaceutical Industry and the Promise of New Biology

1. The definition of a "blockbuster drug" is generally a therapy that achieves peak sales of at least 1 billion USD, with some major blockbuster products achieving over 10 billion USD, such as Pfizer's cholesterol-lowering drug Lipitor. The first global blockbuster drug was Glaxo's Zantac in 1987 and in 2012 there were 116 blockbuster drugs, according to PharmaForum (2012) "Redefining the Blockbuster Model: Why the $1 Billion Entry Point Is No Longer Sufficient–Part 1." *PharmaForum*, September 11, 2012.
2. These interviews (15 in total) were conducted as part of an ESRC Innogen Centre Project called "Innovation Processes in Life Science Industries." They involved very senior R&D mangers and scientists from five of the top ten pharmaceutical firms at the time; and some representatives from smaller biotechnology firms, industry consultants, and analysts. The interviews were focused predominantly on how developments in the life sciences were affecting R&D processes and commercial strategies in these large companies, and what value and future potential biology-based therapies were likely to have relative to conventional small-molecule drugs.
3. The Bayh-Dole Act was enacted on December 12, 1980 (P.L. 96–517, Patent and Trademark Act Amendments of 1980) and for the first time created a uniform patent policy among all the federal agencies that fund research. This enabled public sector organizations, including universities, to retain title to inventions made under federally funded research programs. It opened the doors for universities to patent their inventions and grant exclusive licenses to commercial firms. For an excellent analysis of this Act and its impact on innovation, see Coriat, Benjamin., Orsi, Fabienne., and Weinstein, Oliver. (2004) "Does Biotech Reflect a New Science-Based Innovation Regime?" *Industry and Innovation* 10(3): 231–253.
4. There is of course much debate about the true cost of drug discovery and it is unclear whether figures taken from only the largest multinational pharmaceutical companies are the most useful. Also, costs will vary considerably depending on both selection criteria and accounting

methods. For a detailed and critical analysis of Di Masi's study (Tufts Center for the Study of Drug Development (CSDD), see Light, Donald W. and Warburton, Rebecca. (2011) "Demythologizing the High Costs of Pharmaceutical Research." *Biosocieties* 6: 34–50. For CSDD's open letter of response to Light and Warburton's critique, see http://csdd.tufts.edu/files/uploads/sponsor_csdd_response.pdf (accessed March 2015). For a detailed response by the CSDD authors to previous critiques of their methods, see Di Masi, Joseph A., Hansen, Ronald W., and Grabowski, Henry G. (2005a) "Reply: Extraordinary Claims Require Extraordinary Evidence." *Journal of Health Economics* 24(5): 1034–1044, and Di Masi, Joseph A., Hansen, Ronald W., and Grabowski, Henry G. (2005b). "Reply: Setting the Record Straight on Setting the Record Straight: Response to the Light and Warburton Rejoinder." *Journal of Health Economics* 24(5): 1049–1053.

5. A "me-too" drug is a therapy that is structurally similar and has the same mode of action to others in the same class of products. Its safety and efficacy profile may be slightly different, but it is a minor, incremental innovation.

3   The "Broken Middle" of Health Innovation

1. Some of the material within this chapter was previously published in Mittra, James. (2013a) "Repairing the 'Broken Middle' of the Health Innovation Pathway: Exploring Diverse Practitioner Perspectives on the Emergence and Role of 'Translational Medicine'." *Science and Technology Studies* 26(3): 103–123.

2. In this, and the following chapter, I use data from over 35 in-depth semi-structured interviews with senior academic life scientists, academic clinicians, health service managers, and representatives from industry and the policy/regulatory communities (in both the United States and Europe). These were conducted for a number of different projects I worked on between 2010 and 2013. Furthermore, relevant policy documents and gray literature from the United Kingdom and United States are used to reveal some of the broader sectoral and professional values that have driven TM approaches.

3. "Bench and Bedside" is a term used to distinguish basic science conducted in the laboratory (bench) and the clinical delivery of products to patients (bedside).

4. The linear model that defined the postwar period was based on the work of Vannevar Bush, who had been head of wartime research as the director of the Office of Scientific Research and Development, and whose report *Science, the Endless Frontier* outlined the relationship between basic and applied research and became the foundation for science policy in the United States. For a more detailed discussion, see Crowley, William F. and Gusella, James F. (2009) "Changing Models of Biomedical Research." *Science Translational Medicine* 1(1): 1–5.

5. Interleukin 11 is a synthetic recombinant protein therapy produced using recombinant DNA technology and used for the treatment of thrombocytopenia (abnormally low level of blood platelets) in cancer patients.

6. An alternative to viewing innovation in these linear terms is to consider broader innovation ecosystems and the various feedback loops and contingencies that shape and influence individual value chains, as discussed recently by myself and colleagues in Mastroeni, Michele, Mittra, James, and Tait, Joyce. (2012) *Methodology for the Analysis of Life Science Innovation Systems (ALSIS) and Its Application to Three Case Studies*, TSB Regenerative Medicine Programme: Value Systems and Business Models REALISE project Final Report, May 29, 2012.

7. I am grateful to Sophie Shott, who conducted a pilot study for me in 2011 on the NIH's translational medicine strategy as part of her unpublished honor's dissertation in molecular genetics, University of Edinburgh. In this section, I draw on some of the findings from this work and the many interesting discussions we had.

8. The NIH has its roots in the late eighteenth century, when the Marine Hospital Service (MHS) was established as a predecessor to the US Public Health Services, which established itself in the late nineteenth century and early twentieth century as a laboratory focused on hygiene research and bacteriology. In 1930 this was redesignated the National Institutes of Health by the Ransdell Act and two new NIH buildings were established. Over a number of decades the budget for the NIH was increased significantly by Congress and a number of Institutes and Centers were established for specific research programs.

9. The NIH is the national medical research agency in the United States and provides federal funding for health-related science, as well as developing national policy and health priority setting. It is the largest single source for medical funding in the world and therefore has a significant impact on the development of medical research.

10. For a full list of programs see http://commonfund.nih.gov/initiativeslist/ (accessed March 2015).

11. For an extended analysis of the T1–T3 definition, and how they might be best operationalized, see Barker, Richard W. and Scannell, Jack W. (2015) "The Life Sciences Translational Challenge: The European Perspective." *Therapeutic Innovation and Regulatory Science* 49(3): 415–424.

12. It is important to note that there is disagreement about the meaning and interpretation of biomarkers such as cholesterol, so although they are well established within the clinic, and I describe them as conventional, this is not to assume that their use and validity are uncontested.

13. In terms of imaging biomarkers, the real innovation now is in functional molecular imaging, where physiological processes at the cellular and subcellular levels can be visualized and characterized. This is far more advanced than simple anatomical imaging, which merely

identifies anatomical structures and includes conventional X-ray and ultrasound, etc. For a more in-depth review of imaging biomarkers, see Wang, Jingsong. (2013) "Imaging Biomarkers for Innovative Drug Development." In Mittra, James. and Milne, Christopher-Paul. (eds.) *Translational Medicine: The Future of Therapy?* (Singapore: Pan-Stanford), pp. 163–187.

14. See http://www.biomarkersconsortium.org/ (accessed March 2015) for more information.

4 Organizational Transformations and the Value of Interdisciplinarity

1. Technology Readiness Levels were developed by NASA in the mid-1970s as a measurement tool for the development of new aeronautical technologies. They served as a management tool to reduce uncertainty and risk at various stages of R&D. For a historical review of TRL's over the past 30 years and a rich description of each level, see Mankins, John C. (2009) "Technology Readiness Assessments: A Retrospective." *Acta Astronautica* 65(9–10): 1216–1223.

2. Academically oriented interdisciplinarity has, according to Lyall et al. (2011), a focus on learning rather than expertise and a real intellectual driver to reflect on emerging disciplines. Problem-oriented interdisciplinarity tends to be seen as more pragmatic and issue focused. The authors recognize that some may be uncomfortable with linking interdisciplinarity with "real-world problems," as it may oversimplify the method and give the impression that it is purely instrumental.

3. The clinician-scientist, or physician-scientist, refers to trained medical practitioners who devote a substantial piece of their time to research (either basic or applied) and have training that enables them to combine these two roles. The expectation is that clinician-scientists can contribute to therapeutic development that will most immediately impact on patient care. There is a long history of the clinician-scientist, but from the 1970s there has been a decline in this role, which has led some to describe the clinician-scientist as an endangered species. Part of the decline can be explained by the pressures of contemporary medicine, but part of the problem may also be the hierarchy of credibility.

4. See http://www.ctmm.nl/pro1/general/home.asp (accessed January, 2015) for more information.

5. For a comprehensive overview of different innovation models, and critical analysis of IPR in pre-competitive collaborations, see Bubela, Tania, Fitzgerald, Garret A., and Gold, Richard E. (2012b) "Recalibrating Intellectual Property Rights to Enhance Translational Research Collaborations." *Science Translational Medicine* 4(122): 1–6.

6. See http://www.biomarkersconsortium.org/projects.php (accessed January 2015) for a complete list and descriptions of both completed and active projects.

7. For a much more detailed description and analysis of the TMRC, see Mittra, James. (2013b) "Exploiting Translational Medicine through Public-Private Partnerships: a Case Study of Scotland's Translational Medicine Research Collaboration (TMRC)." In Mittra, James. and Milne, Christopher-Paul. (eds.) *Translational Medicine: The Future of Therapy?* 2013 (Pan-Stanford: Singapore), pp. 213–229. See also Mittra, James (2008) "Impact of the Life Sciences on the Organisation and Management of R&D in Large Pharmaceutical Firms." *International Journal of Biotechnology* 10(5): 416–440.

8. See http://csdd.tufts.edu/news/complete_story/rd_pr_august_2014 (accessed February, 2015).

5 Regulation, Policy, and Governance of Advanced Therapies

1. See http://www.ich.org/ (accessed January 2015) for full details of ICH guidelines.

2. In Europe, the EMA's Orphan Medicinal Product Designation category can be applied to products that treat, prevent, or diagnose life-threatening or chronically debilitating diseases where the prevalence is not more than 5 in 10,000 people, or it must be unlikely that marketing of the medicine would generate sufficient returns to justify the investment needed for its development. Also, the legislation only applies if no other satisfactory method of diagnosis, prevention, or treatment of the condition can be authorized or, if it does, the orphan product must provide significant, additional benefit. Orphan status provides incentives to the innovator, including protocol assistance, technical advice, and market exclusivity. FDA definitions and incentives are similar, but there are some subtle differences, including the definition of orphan disease being a prevalence of less than 200,000. However, there have been moves to harmonize the approaches of these two major regulatory bodies, and it has been possible from 2007 for companies to make a single application through the EMEA/FDA's Common Application Process, in accordance with European Regulation (EC) No 141/2000 of 16 December 1999 and Commission Regulation (EC) No 847/2000, and the United States section 526 of the Federal Food, Drug, and Cosmetic Act (FDCA) (21 U.S.C. 360bb).

3. See http://www.fda.gov/AboutFDA/CentersOffices/OfficeofMedical ProductsandTobacco/CBER/ucm123340.htm (accessed January 2015).

4. Guideline on the Procedure for Accelerated Assessment, Pursuant to Article 14(9) of Regulation (EC) No 726/2004.

5. Commission Regulation (EC) No 507/2006 of 29 March 2006 on the conditional marketing authorization for medicinal products for human use falling within the scope of Regulation (EC) No 726/2004 of the European Parliament and of the Council.

6. Guideline on Procedures for the Granting of a Marketing Authorisation under Exceptional Circumstances, pursuant to Article 14(8) of Regulation (EC) No 726/2004 (EMEA/357981/2005).

7. DG Enterprise (Enterprise and Industry division of the European Commission) defined a human tissue-engineered product as "any autologous or allogeneic product which contains, consists of, or results in engineered human cells or tissues; and has properties for, or is presented as having properties for, the regeneration, repair or replacement of tissue, where the new tissue or cells, in whole or in part, are structurally and functionally analogous to the original tissue being regenerated, repaired or replaced. Engineered means any process whereby human cells or tissues have been substantially manipulated, so their normal/specific physiological functions have been attained. Human tissue-engineered products are derived from living cells or tissues, with the final product containing viable or nonviable cells. They may, for their function, also contain cellular products, biomolecules, and biomaterials (including chemical substances, scaffolds and matrices" (cited in Bock, Anne-Katrin., and Rodriguez-Cerezo, Elelio (2005), "Human Tissue-Engineered Products: Potential Socio-Economic Impacts of a New European regulatory Framework for Authorisation, Supervision and Vigilance." European Commission Joint Research Centre, Technical Report, EUR 21838: 14).

8. Data were collected from five workshops and a small number of interviews collected between 2010 and 2012 as part of the REALISE project funded by the ESRC and TSB, which I worked on with colleagues at the Innogen Institute. This data related to the development of cultured red blood cells for the transfusion market. I also draw on additional data on challenges facing RM from an ESRC-funded project I was involved in, which explored funding gaps for RM in the United Kingdom. This data comprises over 15 interviews and a workshop-based discussion with RM companies, policymakers, funders/investors, and health-care professionals.

6 The Role of Patients and Publics in Health Innovation

1. The concept of "big data" is becoming increasingly popular in health policy discourses. It is often used to describe any initiative that is based on building large data sets and developing new tools to both capture and analyze that data. However, the term is not always well defined and clarified. Technically, big data relates primarily to the establishment of infrastructure to both capture incredibly large sets of data and link that data continuously so that important trends can be observed and used to improve, in the context of health, disease monitoring and health. The establishment of large biobanks, the collection of data from multicentre clinical trials, and large genomics projects that require massive computing power to sequence data are

partly in the realm of big data, but ultimately it is the linking of multiple data points and data sets, which can be continuously updated, where real long-term value is envisaged. The latter still requires significant advances in computing power and the development of new standards to ensure interoperability.

2. CD4 is a glycoprotein found on the surface of many immune cells, including T helper cells, which are a type of white blood cell essential to the immune system. A reduction in T-cells expressing CD4 indicates a high level of HIV infection, so CD4 count can be used as a surrogate biomarker to determine the success of treatment with antiretroviral drugs.

3. For a more in-depth analysis of how the Herceptin case was framed by the media, both in the United Kingdom and Canada, see Abelson, Julia. and Collins, Patricia, A. (2009) "Media Hyping and the 'Herceptin Access Story': An Analysis of Canadian and UK Newspaper Coverage." *Health Policy* 4 (3): 113–128.

4. See http://www.generationscotland.co.uk/ (accessed March 2015) for further information about the project.

5. For more information on the NIH initiative, see http://bd2k.nih.gov/about_bd2k.html#bigdata (accessed March 2015). For more information on the FDA initiatives, see https://open.fda.gov/update/openfda-innovative-initiative-opens-door-to-wealth-of-fda-publicly-available-data/ (accessed March 2015).

Bibliography

Abelson, Julia, and Collins, Patricia, A. (2009) "Media Hyping and the "Herceptin Access Story": An Analysis of Canadian and UK Newspaper Coverage." *Health Policy* 4(3): 113–128.

ABPI. (2013) *Big Data Roadmap.* (London: Association of the British Pharmaceutical Industry).

Abraham, John (2010) "Pharmaceuticalization of Society in Context: Theoretical, Empirical and Health Dimensions." *Sociology* 44(4): 603–622.

Abraham, John and Davis, Courtney. (2007) "Interpellative Sociology of Pharmaceuticals: Problems and Challenges for Innovation and Regulation in the 21st Century." *Technology Analysis and Strategic Management* 19(3): 387–402.

Academy of Medical Sciences. (2007) *Optimizing Stratified Medicine R&D: Addressing Scientific and Economic Issues,* Report of a meeting organized by the Academy of Medical Sciences, Roche and GE Healthcare, Academy of Medical Sciences, London.

Academy of Medical Sciences. (2011) *A New Pathway for the Regulation and Governance of Health Research,* January 2011, available at: http://www. acmedsci.ac.uk/p99puid209.html (accessed December 2014).

Adams, Samantha A. (2011) "Sourcing the Crowd for Health Services Improvement: the Reflexive Patient and 'Share-Your-Experience' Websites" *Social Science and Medicine* 72: 1069–1076.

Adner, Ron and Kapoor, Rahul. (2010) "Value Creation in Innovation Ecosystems: How the Structure of Technological Interdependence Affects Firm Performance in New Technology Generations." *Strategic Management Journal* 31: 306–333.

Allen, Peter, Ramlogan, Ronnie, and Randles, Sally. (2002) "Complex Systems and the Merger Process." *Technology Analysis and Strategic Management* 14(3): 315–329.

Allison, Malorye. (2012) "Reinventing Clinical Trials." *Nature Biotechnology* 30(1): 41–49.

Altar, C. A. (2008) "The Biomarkers Consortium: On the Critical Path of Drug Discovery." *Clinical Pharmacology and Therapeutics* 83(2): 361–364.

Appadurai, Arjun. (ed.) (1988). *The Social Life of Things: Commodities in Cultural Perspective.* (Cambridge: Cambridge University Press).

Arnold, Katie, Coia, Anthony, Saywell, Scott, Smith, Ty, et al. (2002) "Value Drivers in Licensing Deals." *Nature Biotechnology* 20: 1087.

Aronson, J. K, Cohen, A. K, and Lewis, L. D. (2008) "Clinical Pharmacology— Providing Tools and Expertise for Translational Medicine." *British Journal of Clinical Pharmacology* 65(2): 154–157.

Aspinall, Peter J. (2013) "When Is the Use of Race/Ethnicity Appropriate in Risk Assessment Tools for Preconceptual or Antenatal Genetic Screening and How Should It Be Used?" *Sociology* 47(5): 957–975.

Baratt, R. A, Bowens, S. L, McCune, S. K, Johannessen, J. N, and Buckman, S. Y. (2012) "The Critical Path Initiative: Leveraging Collaborations to Enhance Regulatory Science." *Clinical Pharmacology and Therapeutics* 91(3): 380–383.

Barker, Richard W and Scannell, Jack W. (2015) "The Life Sciences Translational Challenge: The European Perspective." *Therapeutic Innovation and Regulatory Science* 49(3): 415–424.

Beckert, Jens and Aspers, Patrik. (eds.) (2011) *The Worth of Goods: Valuation and Pricing in the Economy* (Oxford: Oxford University Press).

Bensaude-Vincent, Bernadette. (2007) "Nanobots and Nanotubes: Two alternative Biomimetic Paradigms of Nanotechnology." In Riskin, Jessica. (ed.) *Genesis Redux: Essays in the History and Philosophy of Artificial Life* (Chicago: Chicago University Press).

Berggren, Christian. (2001) "Mergers, MNES and Innovation—The Need for New Research Approaches." Paper presented at the ESRC Transnational Communities Conference on Multinational Enterprises, Warwick, September 6–8, 2001, p. 14.

Berglund, L and Tarantal, A. (2009) "Strategies for Innovation and Interdisciplinary Translational Research: Removal of Barriers Through the CTSA Mechanism." *Journal of Investigative Medicine* 57(2): 474–476.

Bhatt, Arun. (2010) "Evolution of Clinical Research: A History before and beyond James Lind." *Perspectives in Clinical Research* 1(1): 1–12.

Birch, Kean. (2007) "The *Virtual* Bioeconomy: The 'Failure' of Performativity and the Implications for Bioeconomics." *Scandinavian Journal of Social Theory* 14: 83–89.

Birch, Kean. (2012) "Knowledge, Place and Power: Geographies of Value in the Bioeconomy." *New Genetics and Society* 31(2): 183–201.

Birch, Kean and Tyfield, David. (2012) "Theorizing the Bioeconomy: Biovalue, Biocapital, Bioeconomies or…What?" *Science, Technology and Human Values* 38(3): 299–327.

BMJ. (2008) "Translational Research, Editorial." *British Medical Journal* 337: 863.

BMJ. (2013) "Value-Based Pricing: Can It Work?" Raftery, James. *BMJ* 347: 1–4. DOI: http://dx.doi.org/10.1136/bmj.f5941.

Bock, Anne-Katrin and Rodriguez-Cerezo, Elelio. (2005) "Human Tissue-Engineered Products: Potential Socio-Economic Impacts of a New European Regulatory Framework for Authorisation, Supervision and

Vigilance," European Commission Joint Research Centre, Technical Report, EUR 21838.

Boon, W. P. C, Moors, E. H, Meijer, A, and Schellekens, H. (2010) "Conditional Approval and Approval Under Exceptional Circumstances as Regulatory Instruments for Stimulating Responsible Drug Innovation in Europe." *Clinical Pharmacology and Therapeutics* 88(6): 848–853.

Boon, Wouter and Broekgaarden, Ria. (2010). "The Role of Patient Advocacy Organisations in Neuromuscular Disease R&D—The Case of the Dutch Neuromuscular Disease Association VSN." *Neuromuscular Disorders* 20(2): 148–151.

Borup, Mads, Brown, Nik, Konrad, Kornelia, and Van Lente, Harro. (2006) "The Sociology of Expectations in Science and Technology." *Technology Analysis and Strategic Management* 18(3–4): 285–298.

Bower, Jane D. (2005) "From the 'Rhetoric of Hope' to the 'Patient-Active Paradigm': Strategic Positioning of Pharmaceutical and Biotechnology Companies." *Technology Analysis & Strategic Management* 17(2): 183–204.

Bowker, Geoffrey C and Starr, Susan L. (1999) *Sorting Things Out: Classification and Its Consequences* (Cambridge: MIT Press).

Brody, Howard and Hunt, Linda M. (2006) "BiDil: Asssssing a Race-Based Pharmaceutical." *Annals of Family Medicine* 4(6): 556–560.

Brown, Nik. (2003) "Hope Against Hype—Accountability in Biopasts, Presents and Futures." *Science Studies* 16(2): 3–21.

Brown, Nik. (2013) "Contradictions of Value: Between Use and Exchange in Cord Blood Bioeconomy." *Sociology of Health and Illness* 35(1): 97–112.

Brown, Nik, Rappert, Brian, and Webster, Andrew. (eds.) (2000) *Contested Futures: A Sociology of Prospective Techno-Science* (Aldershot: Ashgate).

Bubela, Tania, Li, Matthew D, Hafez, Mohamed, Bieber, Mark, and Atkins, Harold. (2012a) "Is Belief Larger than Fact: Expectations, Optimism and Reality for Translational Stem Cell Research." *BMC Medicine* 10: 133.

Bubela, Tania, Fitzgerald, Garret A, and Gold, Richard E. (2012b) "Recalibrating Intellectual Property Rights to Enhance Translational Research Collaborations." *Science Translational Medicine* 4(122): 1–6.

Bunnage, Mark E. (2011) "Getting Pharmaceutical R&D Back on Target." *Nature Chemical Biology* 7: 335–339.

Cadigan, Jean R, Lassiter, Dragana, Haldeman, Kaaren, Conlon, Ian, et al. (2013) "Neglected Ethical Issues in Biobank Management: Results from a U.S. study." *Life Sciences, Society and Policy* 9(1): 1–13.

Califf, R and Berglund, L. (2010) "Linking Scientific Discovery and Better Health for the Nation: The First Three Years of the NIH's Clinical and Translational Science Awards." *Academic Medicine* 85(3): 457–462.

Callon, Michel and Rabeharisoa, Vololona. (2008) "The Growing Engagement of Emergent Concerned Groups in Political and Economic Life: Lessons from the French Association of Neuromuscular Disease Patients." *Science, Technology and Human Values* 33(2): 230–261.

Calvert, Jane. (2010) "Systems Biology, Interdisciplinarity and Disciplinary Identity." In Parker, John N, Vermeulen, Niki, and Penders, Bart. (eds.) *Collaboration in the New Life Sciences* (Aldershot: Ashgate), pp. 201–218.

Cambrosio, Alberto. Keating, Peter, and Mogoutov, Andrei. (2004) "Mapping Collaborative Work and Innovation in Biomedicine." *Social Studies of Science* 34(3): 325–364.

CAT. (2010) "Challenges with Advanced Therapy Medicinal Products and How to Meet Them." *Nature Reviews Drug Discovery* 9: 195–201.

Caulfield, Timothy, Burningham, Sarah, Joly, Yann, Master, Zubin, Shabani, Mahsa, et al. (2014) "A Review of the Key Issues Associated with the Commercialization of Biobanks." *Journal of the Law and the Biosciences* 1(1): 94–110. DOI:10.1093/jlb/lst004.

Chalmers, Iain, Bracken, Michael B, Djulbegovic, B, Garattini, Silvio, Grant, Jonathan, et al. (2014) "How to Increase Value and Reduce Waste When Research Priorities Are Set." *The Lancet* 383: 156–164.

Chan, An-Wen, Song, Fujian, Vickers, Andrew, Jefferson, Tom, Dickersin, Kay. et al. (2014) "Increasing Value and Reducing Waste: Addressing Inaccessible Research." *The Lancet* 383: 257–265.

Chataway, Joanna, Tait, Joyce, and Wield, David. (2004) "Understanding Company R&D Strategies in Agro-Biotechnology: Trajectories and Blind Spots." *Research Policy* 33: 1041–1057.

Chen, Haidan and Gottweis, Herbert. (2013) "Stem Cell Treatments in China: Rethinking the Patient Role in the Global Bio-economy." *Bioethics* 27(4): 194–207.

Chiesa, Vittorio and Toletti, Giovanni. (2004) "Network of Collaborations for Innovation: The Case of Biotechnology." *Technology Analysis and Strategic Management* 16(1): 73–76.

Clinical and Translational Science Awards. (2011) *About CTSAs*, available at: http://www.ctsaweb.org/index.cfm?fuseaction=home.aboutHome (accessed November 2014).

Cockburn, Iain M. (2004) "The Changing Structure of the Pharmaceutical Industry." *Health Affairs* 23(1): 10–22.

Cohen, Wesley M and Levinthal, Daniel A. (1990) "Absorptive Capacity: A New Perspective on Learning and Innovation." *Administrative Science Quarterly* 35: 128–152.

Cohen, Joshua, Stolk, Elly, and Niezen, Maartje. (2007) "The Increasingly Complex Fourth Hurdle for Pharmaceuticals." *Pharmacoeconomics* 25(9): 727–734.

Cook, David, Brown, Dearg, Alexander, Robert, March, Ruth, Morgan, Paul, et al. (2014) "Lessons Learned from the Fate of AstraZeneca's Drug Pipeline: A Five-Dimensional Framework." *Nature Reviews Drug Discovery* 13: 419–431.

Cooksey, David. (2006) *A Review of UK Health Research Funding*. UK Department of Health, London, available at: http://www.official-documents.gov.uk/document/other/0118404881/0118404881.pdf (accessed January 2015).

Coombs, Rod and Metcalfe, Stanley J. (2002) "Innovation in Pharmaceuticals: Perspectives on the Co-ordination, Combination and Creation of Capabilities." *Technology Analysis & Strategic Management* 14(3): 261–271.

Cooper, Melinda E. (2008) *Life as Surplus: Biotechnology and Capitalism in the Neoliberal Era* (University of Washington Press: Seattle).

Coriat, Benjamin, Orsi, Fabienne, and Weinstein, Oliver. (2004) "Does Biotech Reflect a New Science-Based Innovation Regime?" *Industry and Innovation* 10(3): 231–253.

Courtney, Aidan, de Sousa, Paul, George, Carol, Laurie, Graeme, and Tait, Joyce. (2011) "Balancing Open Source Stem Cell Science with Commercialization." *Nature Biotechnology* 29(2): 115–116.

Cox, Helen and Webster, Andrew. (2012) "Translating Biomedical Science into Clinical Practice: Molecular Diagnostics and the Determination of Malignancy." *Health* 17(4): 391–406.

Crowley, William F and Gusella, James F. (2009) "Changing Models of Biomedical Research." *Science Translational Medicine* 1(1): 1–5.

CTMM. (2006) *Business Plan: Center for Translational Molecular Medicine*, May 2006 http://www.ctmm.nl/en/downloads/about-ctmm_downloads/business-plan (accessed November 2014).

Cuende, Natividad, Boniface, Christelle, Bravery, Christopher, Forte, Miguel, Giordano, Rosaria, et al. (2014) "The Puzzling Situation of Hospital Exemption for Advanced Therapy Medicinal Products in Europe and Stakeholders' Concerns'." *Cytotherapy* 16: 1597–1600.

Cutler, David M and McClellan, Mark. (2001) "Is Technological Change in Medicine Worth It?" *Health Affairs* 20(5): 11–29.

Cyranoski, David. (2013) "Japan to Offer Fast-Track Approval Path for Stem Cell Therapies." *Nature Medicine* 19: 510.

Dahl, Svein G and Sylte, Ingebrigt. (2006) "From Genomics to Drug Targets." *Journal of Psychopharmacology* 20(4): 95–99.

Danzon, Patricia M, Epstein, Andrew, and Nicholson, Sean. (2004) "Mergers and Acquisitions in the Pharmaceutical and Biotech Industries," NBER Working Paper Series, no. 10536, 2004, National Bureau of Economic Research, Cambridge, MA.

Davies, Gail. (2010) "Captivating Behaviour: Mouse Models, Experimental Genetics and Reductionist Returns in the Neurosciences." In Parry, Sarah and Dupre, John. (eds.) *Nature after the Genome* (Wiley Blackwell: London), pp. 53–72.

Davies, Gail, Frow, Emma, and Leonelli, Sabina. (2013) "Bigger, Faster, Better? Rhetorics and Practices of Large-Scale Research in Contemporary Bioscience." *Biosocieties* 8: 386–396.

DiMaggio, Paul and Powell, Walter W. (1983) "The Iron Cage Revisited: Institutional Isomorphism and Collective Rationality in Organizational Fields." *American Sociological Review* 48: 147–160.

DiMasi, Joseph A, Hanson, Ronald W, and Grabowski, Henry G. (2003) "The Price of Innovation: New Estimates of Drug Development Costs." *Journal of Health Economics* 22: 151–185.

Di Masi, Joseph A, Hansen, Ronald W, and Grabowski, Henry G. (2005a) "Reply: Extraordinary Claims Require Extraordinary Evidence." *Journal of Health Economics* 24(5): 1034–1044.

Di Masi, Joseph A, Hansen, Ronald W, and Grabowski, Henry G. (2005b) "Reply: Setting the Record Straight on Setting the Record Straight: Response to the Light and Warburton Rejoinder." *Journal of Health Economics* 24(5): 1049–1053.

Dougherty, Denise and Conway, Patrick H. (2008) "The 3Ts Roadmap to Transform US Healthcare: The 'How' of High-Quality Care." *JAMA* 299: 2319–2321.

Dove, Alan. (2003) "Walking the Drug Regulatory Tightrope." *Nature Biotechnology* 21: 495–498.

Drews, Jürgen. (2000) "Drug Discovery: A Historical Perspective." *Science* 287: 1960–1964.

Drews, Jürgen and Ryser, Stefan. (1996) "Innovation Deficit in the Pharmaceutical Industry." *Drug Information Journal* 30: 97–108.

Drolet, Brian C and Lorenzi, Nancy M. (2010) "Translational Research: Understanding the Continuum from Bench to Bedside." *Translational Research* 157(1): 1–5.

Durst, Susanne and Poutanen, Petro. (2013) "Success Factors of Innovation Ecosystems—Initial Insights from a Literature Review." In Smeds, Riita and Irrmann Olivier. (eds.) *Co-Create 2013: The Boundary-Crossing Conference on Co-design in Innovation, Conference Proceedings*, Aalto University Publication Series, pp. 27–38.

Dupré, John. (1995) *The Disorder of Things: Metaphysical Foundations of the Disunity of Science* (Cambridge: Harvard University Press).

Dussuage, Isabelle, Helgesson, Claes-Fredrik, and Lee, Francis. (eds.). (2015) *Value Practices in the Life Sciences and Medicine* (Oxford: Oxford University Press).

Eger, Stephan and Mahlich, Jörg C. (2014) "Pharmaceutical Regulation in Europe and Its Impact on Corporate R&D." *Health Economics Review* 4(23): 1–9.

Ehmann, F, Amati, M Papaluca, Salmonson, T, Posch, M, Vamvakas, S, et al. (2013) "Gatekeepers and Enablers: How Drug Regulators Respond to a Challenging and Changing Environment by Moving toward a Proactive Attitude." *Clinical Pharmacology and Therapeutics* 93(5): 425–432.

Eichler, H.-G. Oye, K, Baird, L. G, Abadie, E, Brown, J, et al. (2012) "Adaptive Licensing: Taking the Next Step in the Evolution of Drug Approval." *Clinical Pharmacology and Therapeutics* 91(3): 426–437.

Eichler, H.-G, Baird, L. G, Barker, R, Bloechl-Daum, B, Børlum-Kristensen, F, et al. (2015) "From Adaptive Licensing to Adaptive Pathways: Delivering a Flexible Life-Span Approach to Bring New Drugs to Patients." *Clinical Pharmacology and Therapeutics* 97(3): 234–246.

EMA. (2011) *Road Map to 2015: The European Medicines Agency's Contribution to Science, Medicines and Health*, European Medicines Agency, London.

EMA. (2014a) "European Medicines Agency Launches Adaptive Licensing Pilot Project." European Medicines Agency Press release, March 19, 2014, EMA/430892/2013.

EMA. (2014b) *Reflection Paper on the Use of Patient Reported Outcome (PRO) Measures in Oncology Studies,* Draft EMA/CHMP/292464/2014, European Medicines Agency, London.

Epstein, Steven. (1996) *Impure Science: AIDS, Activism and the Politics of Knowledge* (Oakland: UC Press).

Epstein, Steven. (2007) *Inclusion: The Politics of Difference in Medical Research,* (Chicago: University of Chicago Press).

Espeland, Wendy Nelsoon and. & Stevens, Mitchell, L. (1998) "Commensuration as a Social Process." *Annual Review of Sociology* 24: 313–343.

Etzkowitz, Henry. (2006) "The New Visible Hand: An Assisted Linear Model of Science and Innovation Policy." *Science and Public Policy* 33(5): 310–320.

Etzkowitz, Henry. (2008) *The Triple Helix: University-Industry-Government Innovation in Action* (London: Routledge).

European Risk Forum. (2014) "The Innovation Principle: Stimulating Innovation, Jobs and Growth," An open letter to Mr Jean-Claude Juncker, November 4, 2014, available at: http://www.riskforum.eu/uploads/2/5/7/1/25710097/innovation_principle_letter_4_nov.pdf (accessed January 2015).

European Parliament and Council of the European Union. (2007) "Regulation (EC) No 1394/2007 of the European Parliament and of the Council of November 13, 2007 on Advanced Therapy Medicinal Products and Amending Directive 2001/83/EC and Regulation (EC) No 726/2004." *Official Journal* L 324: 121–137, December 10.

Le Fanu, James. (2011) *The Rise and Fall of Modern Medicine* (London: Abacus).

Faulkner, Alex. (2012) "Laws's Performativities: Shaping the Emergence of Regenerative Medicine through European Union legislation." *Social Studies of Science* 42(5): 753–774.

FDA. (2004) *Innovation or Stagnation: Challenges and Opportunities on the Critical Path to New Medical Products,* March 2004, available at: http://www.fda.gov/ScienceResearch/SpecialTopics/CriticalPathInitiative/CriticalPathOpportunitiesReports/ucm077262.htm (accessed May 2014).

FDA. (2006) *Critical Path Opportunities Report,* US Department of Health and Human Services, March, 2006, available at http://www.fda.gov/downloads/ScienceResearch/SpecialTopics/CriticalPathInitiative/CriticalPathOpportunitiesReports/UCM077254.pdf (accessed May 2014).

FDA. (2009) *Guidance for Industry: Patient-Reported Outcome Measures: Use in Medical Product Development to Support Labeling Claims,* US Department of Health and Human Services, December 2009, available at: http://www.fda.gov/downloads/Drugs/Guidances/UCM193282.pdf (accessed May 2014).

FDA. (2011) *Advancing Regulatory Science at the FDA: A Strategic Plan*, August 2011, available at http://www.fda.gov/downloads/ ScienceResearch/SpecialTopics/RegulatoryScience/UCM268225.pdf (accessed May 2014).

FDA. (2012) *Paving the Way for Personalized Medicine: FDA's Role in a New Era of Product Development*, US Department of Health and Human Services, October 2013, available at: http://www.fda.gov/downloads/ ScienceResearch/SpecialTopics/PersonalizedMedicine/UCM372421. pdf (accessed March 2015).

Featherstone, James and Renfrey, Siân. (2004) "The Licensing Gamble: Raising the Stakes." *Nature Biotechnology* 3: 107–108.

Feldman, Bonnie, Martin, Ellen M, and Skotnes, Tobi. (2012) *Big Data in Healthcare Hype and Hope*, Dr. Bonnie 360, Business Development for Digital Health, October 2012, available at: http://www.west-info.eu/ files/big-data-in-healthcare.pdf (accessed December 2014).

Ferrara, Joseph. (2007) "Personalized Medicine: Challenging Pharmaceutical and Diagnostic Company Business Models." *McGill Journal of Medicine* 10: 59–61.

Caron-Flinterman, Francisca J, Broerse, Jaqueline E. W, and Bunders, Joske F. G. (2005) "The Experiential Knowledge of Patients: A New Resource for Biomedical Research?" *Social Science and Medicine* 60: 2575–2584.

Fourcade, Marion. (2011a) "Price and Prejudice: On Economics and the Enchantment (and Disenchantment) of Nature." In Beckert, Jens and Aspers, Patrik. (eds.) (2011) *The Worth of Goods: Valuation and Pricing in the Economy* (Oxford: Oxford University Press), pp. 41–62.

Fourcade, Marion. (2011b) "Cents and Sensibility: Economic Valuation and the Nature of "Nature"." *American Journal of Sociology*, 116(6): 1721–1777.

Frow, Emma. (2013) "Making Big Promises Come True? Articulating and Realizing Value in Synthetic Biology." *Biosocieties* 8: 432–448.

Fuller, Steve and Lipinska, Veronika. (2014) *The Proactionary Imperative: A Foundation for Transhumanism* (Basingstoke: Palgrave Macmillan).

Garcia-Sancho, Miguel. (2012) *Biology, Computing and the History of Molecular Sequencing: From Proteins to DNA, 1945–2000* (Basingstoke: Palgrave Macmillan).

Garavaglia, Christian, Malerba, Franco, and Orsenigo, Luigi. (2006) "Entry, Market Structure, and Innovation in a 'History-Friendly' Model of the Evolution of the Pharmaceutical Industry." In Mazzucato, Mariana and Dosi, Giovanni. (eds.) *Knowledge Accumulation and Industry Evolution: The Case of Pharma-Biotech* (Cambridge: Cambridge University Press), pp. 234–265.

Gibbons, Michael, Nowotny, Helga, and Limoges, Camille. (1994) *The New Production of Knowledge: The Dynamics of Science and Research in Contemporary Societies* (London: Sage).

Glasziou, Paul, Altman, Douglas G, Bossuyt, Patrick, Boutron, Isabelle, et al. (2014) "Reducing Waste from Incomplete or Unusable Reports of Biomedical Research." *The Lancet* 383: 267–275.

Gold, Richard E, Herder, Matthew, and Trommetter, Michel. (2007) *The Role of Biotechnology Intellectual Property Rights in the Bioeconomy of 2030,* OECD International Futures Project on "The Bioeconomy to 2030: Designing a Policy Agenda," December 2007, OECD, available at: http://www.iwbio.de/fileadmin/Publikationen/Englisch/OECD_12-2007.pdf (accessed February 2015).

Goldman, Dana P, Gupta, Charu, Vasudeva, Eshan, Trakas, Kostas, Riley, Ralph, et al. (2013) "The Value of Diagnostic Testing in Personalized Medicine." *Forum for Health Economics and Policy* 16(2): 121–133.

Goven, Joanna and Pavone, Vincenzo. (2015) "The Bioeconomy as Political Project: A Polanyian Analysis." *Science, Technology and Human Values* 40(3): 302–337.

Grabowski, Henry. (2011) "The Evolution of the Pharmaceutical Industry over the Past 50 Years: A Personal Reflection." *International Journal of the Economics of Business* 18(2): 161–176.

Graves, Samuel B and Langowitz, Nan S. (1993) "Innovative Productivity and Returns to Scale in the Pharmaceutical Industry." *Strategic Management Journal* 14: 593–695.

Greenhalgh, Trisha, Howick, Jeremy, and Maskrey, Neal. (2014) "Evidence Based Medicine: A Movement in Crisis?" *British Medical Journal* 348: 1–7. DOI: 10.1136/bmj.g3725.

Haddow, Gillian, Laurie, Graeme, Cunningham-Burley, Sarah, and Hunter, Kathryn G. (2007) "Tackling Community Concerns about Commercialisation and Genetic Research: A Modest Interdisciplinary Proposal." *Social Science and Medicine* 64: 272–282.

Hallowell, Nina, Heiniger, Louise, Baylock, Brandi, Price, Melanie, Butow, Phyllis, et al. (2015) "Rehabilitating the Sick Role: The Experiences of High-Risk Women Who Undergo Risk Reducing Breast Surgery." *Health Sociology Review.* 24(2): 186–198. DOI: 10.1080/14461242.2014.999402.

Hamburg, Margaret A. (2010) "Innovation, Regulation, and the FDA." *The New England Journal of Medicine* 363(23): 2228–2232.

Hamilton, Chris. (2008) "Intellectual Property Rights, the Bioeconomy and the Challenge of Biopiracy." *Genomics, Society and Policy* 4(3): 26–45.

Hanlin, Rebecca, Chataway, Joanna, and Smith, James. (2007) "Global Health Public-Private Partnerships: IAVI, Partnerships and Capacity Building." *African Journal of Medical Science* 36: 69–75.

Hara, Takuji. (2003) *Innovation in the Pharmaceutical Industry: The Process of Drug Discovery and Development* (London: Edward Elgar).

Hartley, Keith and Maynard, Alan. (1982) "The Regulation of the UK Pharmaceutical Industry: A Cost-Benefit Analysis." *Managerial and Decision Economics* 3(3): 122–130.

Hedgecoe, Adam. (2004) *The Politics of Personalized Medicine: Pharmacogenetics in the Clinic* (Cambridge: Cambridge University Press).

Helgesson, Claes-Fredrik and Kjellberg, Hans. (2013) "Introduction: Values and Valuations in Market Practice." *Journal of Cultural Economy* 6(4): 361–369.

Helgesson, Claes-Fredrik and Muniesa, Fabian. (2013) "For What It's Worth: An Introduction to Valuation Studies." *Valuation Studies* 1(1): 1–10.

Helmreich, Stefan. (2008) "Species of Biocapital." *Science as Culture* 17(4): 463–478.

Henderson, Rebecca M. (2000) "Drug Industry Mergers Won't Necessarily Benefit R&D." *Research Technology Management* 43(4):10–11.

Henderson, Rebecca and Cockburn, Iain. (1997) "Firm Size and Research Productivity in Drug Discovery," Working Paper; 10–11, available at: http://www.mit.edu/afs/athena/course/15/15.141/readings/insee97.pdf (accessed March 2014).

Henderson, Gail E, Cadigan, Jean R, Edwards, Teresa P, et al. (2013) "Characterizing Biobanks Organizations in the U.S: Results from a National Survey." *Genome Medicine* 5(3): 1–12.

Herxheimer, Andrew. (2003) "Relationships between the Pharmaceutical Industry and Patients' Organisations." *British Medical Journal* 326: 1208–1210.

Heuts, Frank and Mol, Annemarie. (2013) "What Is a Good Tomato? A Case of Valuing in Practice." *Valuation Studies* 1(2): 125–146.

Hilgartner, Stephen. (2007) "Making the Bioeconomy Measurable: Politics of an Emerging Anticipatory Machinery." *Biosocieties* 2(3): 382–386.

Hogarth, Stuart. (2015) "Neoliberal Technocracy: Explaining How and Why the US Food and Drug Administration Has Championed Pharmacogenomics." *Social Science and Medicine* 131: 255–262.

HOL. (2013) *Regenerative Medicine Report,* Science and Technology Committee, 1st Report of Session 2013–2014, London, the Stationary Office Ltd, available at: http://www.publications.parliament.uk/pa/ld201314/ldselect/ldsctech/23/23.pdf (accessed January 2014).

Holland-Moritz, Pamela. (2006) "Moving Biotech Products from CBER to CDER: A Work in Progress." *Pharmaceutical Regulatory Guidance Book* July, 2006: 70–73.

Holmberg, Tora, Schwennesen, Nete, and Webster, Andrew. (2011) "Bio-Objects and the Bio-Objectification Process." *Croatian Medical Journal* 52(6): 740–742.

van den Hoonaard, Deborah K. (2009) "Moving toward a Three-Way Intersection in Translational Research: A Sociological Perspective." *Qualitative Health Research* 19(12):1783–1787.

Hopkins, Michael. (2006) "The Hidden Research System: The Evolution of Cytogenetic Testing in the National Health Service." *Science as Culture* 15(3): 253–276.

Hopkins, Michael M, Martin, Paul A, Nightingale, Paul, Kraft, Alison, and Mahdi, Surya. (2007) "The Myth of the Biotech Revolution: An Assessment of Technological, Clinical and Organisational Change." *Research Policy* 36(4): 566–589.

Hopkins, Michael M, Crane, Philippa A, Nightingale, Paul, and Baden-Fuller, Charles. (2013) "Buying Big into Biotech: Scale, Financing, and the Industrial Dynamics of UK biotech, 1980–2009." *Industrial and Corporate Change* 22(4) 903–952.

Horig, Heidi and Pullman, William. (2004) "From Bench to Clinic and Back: Perspective on the First 1st IQPC Translational Research Conference." *Journal of Translational Medicine* 2(44): 1–8. DOI: 10.1186/1479-5876-2-44.

Horrobin, David F. (2001) "Realism in Drug Discovery—Could Cassandra Be Right?" *Nature Biotechnology* 19: 1099–1100.

House of Commons. (2005) *The Influence of the Pharmaceutical Industry*, Health Committee 4th Report of Session 2004–2005, April 5, 2005, The Stationary Office Ltd, London, available at: http://www.lindalliance.org/pdfs/HofCHealthCommittee.pdf (accessed April 2013).

Howells, Jeremy. (2002) "Mind the Gap: Information and Communication Technologies, Knowledge Activities and Innovation in the Pharmaceutical Industry." *Technology Analysis and Strategic Management* 14(3): 356–367.

Huber, Brian and Doyle, John. (2010) *Oncology Drug Development and Value-Based Medicine*, Quintiles White Paper, available at: http://www.quintiles.com/~/media/library/white%20papers/oncology-drug-development-and-value-based-medicine.pdf (accessed December 2014).

Huckman, Robert S and Strick, Eli. (2005) "GlaxoSmithKline: Reorganizing Drug Discovery" (A). *Harvard Business School Case Study*, 9–605–074, May 2005.

Huzair, Farah and Papaioannou, Theo. (2012) "UK Biobank: Consequences for Commons and Innovation." *Science and Public Policy* 39(4): 500–512.

Kaitin, K. I. (2010) "Deconstructing the Drug Development Process: The New Face of Innovation." *Clinical Pharmacology and Therapeutics* 87: 469–479.

Kayyali, Basel, Knott, David, and Van Kuiken, Steve. (2013) *The Big-Data Revolution in US Healthcare: Accelerating Value and Innovation*, Mckinsey & Company Report, available at: http://www.mckinsey.com/insights/health_systems_and_services/the_big-data_revolution_in_us_health_care (accessed January 2015).

Kelly, Ann H and Geissler, P Wenzel. (2011) "The Value of Transnational Medical Research." *Journal of Cultural Economy* 4(1): 3–10.

Kent, Julie, Faulkner, Alex, Geesink, Ingrid, and Fitzpatrick, David. (2006) "Towards Governance of Human Tissue Engineered Technologies in Europe: Framing the Case for a New Regulatory Regime." *Technological Forecasting and Social Change* 73: 41–60.

Klein, Julie Thompson. (2000) "A Conceptual Vocabulary of Interdisciplinary Science." In Weingart, Peter and Stehr, Nico. (eds.) *Practicing Interdisciplinarity* (Toronto: University of Toronto Press), pp. 3–24

Klein, Julie Thompson. (2004) "Interdisciplinarity and Complexity: An Evolving Relationship." *E: CO Special Double Issue* 6(1–2): 2–10.

Kling, Jim. (2006) "Careers in Systems Biology: Working the Systems." *Science* 311: 1305–1306.

Kooijman, Marlous. (2013) *Why Animal Studies Are Still Being Used in Drug Development: An Innovation Systems Perspective*, PhD Thesis, ISBN 978–90–6464–735–2.

Korenstein, Deborah. (2015) "Patient Perception of Benefits and Harms: The Achilles Heel of High-Value Care." *JAMA Internal Medicine* 175(2): 287–288.

Kraft, Alison. (2004) "Genomics as an Emergent Industry: Panacea for Pharmaceutical Innovation?" Paper presented at Business and Society Conference, June 2004.

Kraft, Alison. (2013) "New Light through an Old Window: The Translational Turn in Biomedical Research—a Historical Perspective." In Mittra, James and Milne, Christopher-Paul. (eds.) *Translational Medicine: The Future of Therapy?* (Singapore: Pan-Stanford), pp. 19–53.

Kraft, Alison and Rothman, Harry. (2008) "Genomics-Based Drug Innovation: Visions and Commercial Viability." *International Journal of Biotechnology* 10(5): 441–460.

Lam, Michael D. (2004) "Dangerous Liaisons." *Pharmaceutical Executive* 24(5): 1–3, avaialble at: http://www.pharmexec.com/node/241803?rel=canonical (accessed March 2015).

Lamont, Michèle. (2012) "Toward a Comparative Sociology of Valuation and Evaluation." *Annual Review of Sociology* 38: 201–221.

Langstrup, Henriette. (2011) "Interpellating Patients as Users: Patient Associations and the Project-Ness of Stem Cell Research." *Science, Technology and Human Values* 36(4): 573–594.

Ledford, Heidi. (2008) "Translational Research: The Full Cycle." *Nature* 453: 843–845.

Lewin, Benjamin. (2015) "Health Care Collaboration Between Patients and Physicians." In Penders, Bart, Vermeulen, Niki, and Parker, John N. (eds.) *Collaboration across Health Research and Medical Care* (Surrey: Ashgate), pp. 195–214.

Lewis, Jamie, Atkinson, Paul, Harrington, Jean, and Featherstone, Katie. (2013) "Representation and Practical Accomplishment in the Laboratory: When Is an Animal Model Good-Enough?" *Sociology* 47(4): 776–792.

Lezaun, Javier and Montgomery, Catherine M. (2014) "The Pharmaceutical Commons: Sharing and Exclusion in Global Health Drug Development." *Science, Technology and Human Values* 40(1): 3–29.

Light, Donald W and Warburton, Rebecca. (2011) "Demythologizing the High Costs of Pharmaceutical Research." *Biosocieties* 6: 34–50.

Lindblom, Charles E. (1959) "The Science of 'Muddling Through'." *Public Administration Review* 19(2): 79–88.

Lowy, Ilana. (1997) *Between Bench and Bedside. Science, Healing and Interleukin-2 in a Cancer Ward* (Cambridge: Harvard University Press).

Lowy, Ilana. (2011) "Historiography of Biomedicine: 'Bio,' 'Medicine,' and in between." *ISIS* 102(1): 116–122.

Lukkonen, Terttu. (2005) "Variability in Organisational Forms of Biotechnology Firms." *Research Policy* 34(4): 555–570.

Lyall, Catherine, Bruce, Ann, Tait, Joyce, and Meagher, Laura. (2011) *Interdisciplinary Research Journeys: Practical Strategies for Capturing Creativity* (London: Bloomsbury).

Lyall, Catherine, Bruce, Ann, Marsden, Wendy, and Meagher, Laura. (2013) "The Role of Funding Agencies in Creating Interdisciplinary Knowledge." *Science and Public Policy* 40: 62–71.

Maciulaitis, Romaldas, D'Apote, Lucia, Buchanan, Andrew, Pioppo, Laura, and Schneider, Christian K. (2012) "Clinical Development of Advanced Therapy Medicinal Products in Europe: Evidence That Regulators Must be Proactive." *The American Journal of Gene and Cell Therapy* 20(3): 479–482.

Malerba, Franco and Orsenigo, Luigi. (2002) "Innovation and Market Structure in the Dynamics of the Pharmaceutical Industry and Biotechnology: Towards a History-Friendly Model." *Industrial and Corporate Change* 11(4): 667–703.

Malik, Nafees N. (2014) "Reimbursement and Adoption of Advanced Therapies: The %-C Framework." *Regenerative Medicine* 9(5): 573–578.

Mankins, John C. (2009) "Technology Readiness Assessments: A Retrospective." *Acta Astronautica*, 65 (9–10): 1216–1223.

Mankoff, Stacey P, Brander, Christian, Ferrone, Soldano, and and Marincola, Francesco, M. (2004) "Lost in Translation: Obstacles to Translational Medicine." *Journal of Translational Medicine* 2(14). DOI:10.1186/1479–5876–2–14.

Martin, Paul, Brown, Nik, and and Kraft, Alison. (2008) "From Bedside to Bench? Communities of Promise, Translational Research and the Making of Blood Stem Cells." *Science as Culture* 17(1): 29–41.

Mastroeni, Michele, Mittra, James, and Tait, Joyce. (2012) *Methodology for the Analysis of Life Science Innovation Systems (ALSIS) and Its Application to Three Case Studies*, TSB Regenerative Medicine Programme: Value Systems and Business Models REALISE project Final Report, May 29, 2012, available at: http://www.genomicsnetwork.ac.uk/media/REALISE%20Case%20Study%20Report-Innogen.pdf (accessed March 2014).

Mavris, M and Le Cram, Y. (2012) "Involvement of Patient Organisations in Research and Development of Orphan Drugs for Rare Diseases in Europe." *Molecular Syndromology* 3(5): 237–243.

Mazanderani, Fadhila. (2014) "The Patient's View: Perspectives from Neurology and the 'New' Genetics." *Science as Culture* 23(1): 135–144.

Mazanderani, Fadhila, Lockock, Louise, and Powell, John. (2013) "Biographical Value: Towards a Conceptualisation of the Commodification of Illness Narratives in Contemporary Healthcare." *Sociology of Health and Illness* 35(6): 891–905.

Mazzucato, Mariana. (2013) *The Entrepreneurial State: Debunking Public vs. Private Sector Myths.* (London: Anthem Press).

McCarthy, John. (2004) "Tackling the Challenges of Interdisciplinary Bioscience." *Nature Reviews Molecular Cell Biology* 5: 933–937.

McMahon, Aisling and Harmon, Shawn, H. E. (2012) "Banking (on) the Brain: A Report on the Legal and Regulatory Concerns." *SCRIPTed* 9(3): 376–383. DOI:10.2966/scrip.090312.376.

Meadowcroft, John. (2008) "Patients, Politics, and Power: Government Failure and the Politicization of U.K. Health Care." *Journal of Medicine and Philosophy* 33: 427–444.

van Meer, P. J. K, Kooijman, M, Van der Laan, J. W, Moors, H. E. M, and Schellekens, H. (2013) "The Value of Non-Human Primates in the Development of Monoclonal Antibodies." *Nature Biotechnology* 31: 882–883.

van Meer, Peter J. K. Ebbers, Hans C, Kooijman, Marlous, Gispen-de-Wied, Christine C, Silva-Lima, Beatriz, et al. (2015) "Contribution of Animal Studies to Evaluate the Similarity of Biosimilars to Reference Products." *Drug Discovery Today* 20(4): 483–490. DOI:10.1016/j.drudis.2014.11.009.

Meltzer, Ingrid and Webster, Andrew. (2011) "Bio-Objects and Their Boundaries: Governing Matters at the Intersection of Society, Politics and Science." *Croation Medical Journal* 52(5): 648–650.

Messner, Donna. (2008) *Fast Track: The Practice of Drug Development and Regulatory Innovation in the Late Twentieth Century US,* Unpublished PhD Thesis, University of Edinburgh.

Messner, D. A and Tunis, S. R. (2012) "Current and Future State of FDA-CMS Parallel Reviews." *Clinical Pharmacology and Therapeutics* 91(3): 383–385.

MHRA (2014) *Rules and Guidance for Pharmaceutical Manufacturers and Distributors,* MHRA, 2 January 2014, Crown Copyright: http://www.mhra.gov.uk/Publications/Regulatoryguidance/Medicines/CON2030291 (accessed December 2014).

Milne, Christopher-Paul. (2009) "Can Translational Medicine Bring Us out of the R&D Wilderness?" *Personalized Medicine* 6(5): 543–553.

Milne, Christopher-Paul. (2013) "Translational Medicine: The Industry Perspective." In Mittra, James and Milne, Christopher-Paul. (eds.) *Translational Medicine: The Future of Therapy?* (Singapore: Pan-Stanford), pp. 55–80.

Milne, Christopher-Paul and Tait, Joyce. (2009) "Evolution along the Government-Governance Continuum: FDA's Orphan Products and Fast Track Programs as Exemplars of "What Works' for Innovation and Regulation." *Food and Drug Law Journal* 64(4): 733–753.

Mittra, James. (2006) "The Socio-Political Economy of Pharmaceutical Mergers: A Case study of Sanofi and Aventis." *Technology Analysis and Strategic Management* 18(5): 473–496.

Mittra, James. (2007a) "Predictive Genetic Information and Access to Life Assurance: The Poverty of Genetic Exceptionalism." *Biosocieties* 2(3): 349–373.

Mittra, James. (2007b) "Life Science Innovation and the Re-Structuring of the Pharmaceutical Industry: Mergers, Acquisitions and Strategic Alliances." *Technology Analysis and Strategic Management* 19(3): 279–301.

Mittra, James. (2008) "Impact of the Life Sciences on the Organisation and Management of R&D in Large Pharmaceutical Firms." *International Journal of Biotechnology* 10(5): 416–440.

Mittra, James. (2013a) "Repairing the 'Broken Middle' of the Health Innovation Pathway: Exploring Diverse Practitioner Perspectives on the Emergence and Role of "Translational Medicine." *Science and Technology Studies* 26(3): 103–123.

Mittra, James. (2013b) "Exploiting Translational Medicine through Public-Private Partnerships: a Case Study of Scotland's Translational Medicine Research Collaboration (TMRC)." In Mittra, James and Milne, Christopher-Paul. (eds.) *Translational Medicine: The Future of Therapy?* (Singapore: Pan-Stanford), pp. 213–229.

Mittra, James, Tait, Joyce, and Wield, David. (2011) "From Maturity to Value Added Innovation: Lessons from the Pharmaceutical and Agro-Biotechnology Industries." *Trends in Biotechnology* 29(3): 105–109.

Mittra, James and Milne, Christopher-Paul. (2013) "Introduction to Translational Medicine." In Mittra, James and Milne, Christopher-Paul. (eds.) *Translational Medicine: The Future of Therapy?* (Singapore: Pan-Stanford), pp. 3–16.

Mittra, James and Tait, Joyce. (2012) "Analysing Stratified Medicine Business Models and Value Systems: Innovation-Regulation Interactions." *New Biotechnology* 29(6): 709–719.

Mittra. J, Tait, J, Mastroeni, M, Turner, M. L, Mountford, J. C, and Bruce, Kevin. (2015) "Identifying Viable Regulatory and Innovation Pathways for Regenerative Medicine: A Case Study of Cultured Red Blood Cells." *New Biotechnology* 32(1): 180–190.

Mol, Annemarie. (2003) *The Body Multiple: Ontology in Medical Practice* (Durham: Duke University Press).

Moreira, Tiago, May, Carl, and Bond, John. (2009) "Regulatory Objectivity in Action: Mild Cognitive Impairment and the Collective Production of Uncertainty." *Social Studies of Science* 39(5): 665–690.

Moreira, Tiago, O'Donovan, Orla, and Howlett, Etaoine. (2014) "Assembling Dementia Care: Patient Organisations and Social Research." *Biosocieties* 9(2): 173–193.

Morrison, Michael and Cornips, Lucas. (2012) "Exploring the Role of Dedicated Online Biotechnology News Providers in the Innovation Economy." *Science, Technology and Human Values* 37(3): 262–285.

MRC. (2007) "Translational Medicine." Medical Research Council, London, available at: http://www.mrc.ac.uk/Utilities/Documentrecord/index.htm?d=MRC003561 (accessed January 2012).

MRC. (2008) Translational Research Strategy, Medical Research Council, London, available at: http://www.mrc.ac.uk/consumption/groups/public/documents/content/mrc004551.pdf (accessed January 2012).

MRC. (2013) *Research Changes Lives 2014–2018: Strategic Plan*, Medical Research Council, London, available at: http://www.mrc.ac.uk/news-events/publications/strategic-plan-2014-19/ (accessed January 2014).

Mugwagwa, Julius, Hanlin, Rebecca, Chataway, Joanna, and Muraguri, Lois. (2013) "The Role of the Product Development Partnership as a Translational Mechanism for Delivering Health Solutions in Low-Resource Settings." In Mittra, James and Milne, Christopher-Paul. (eds.) *Translational Medicine: The Future of Therapy?* (Singapore: Pan-Stanford), pp. 233–256.

Narayan, Vaibhav A, Mohwinckel, Marco, Pisano, Gary, Yang, Michael, and Manji, Husseini K. (2013) "Beyond Magic Bullets: True Innovation in Healthcare." *Nature Reviews Drug Discovery* 12: 85–86.

National Center for Research Resources. (2009) *Clinical and Translational Science Awards Progress Report 2006–2008.* National Institutes of Health, US Department of Health and Human Services, available at: http://www.ncats.nih.gov/files/2008_ctsa_progress_report.pdf (accessed March 2015).

Nature. (2008) "Translational Research: Crossing the Valley of Death," Declan Butler, Editorial. *Nature* 453: 840–842.

Naylor, Stephen and Cole, Toby. (2010) "Companion Diagnostics in the Pharmaceutical Industry Part 11: Business Models." *Drug Discovery World* Summer: 61–68.

Nelson, Richard. R and Sampat, Bhaven N. (2001) "Making Sense of Institutions as a Factor Shaping Economic Performance." *Journal of Economic Behavior and Organization* 44: 31–54.

Nelson, Aaron L, Dhimolea, Eugen, and Reichert, Janice M. (2010) "Development Trends for Human Monoclonal Antibody Therapeutics." *Nature Reviews Drug Discovery* 9: 767–774.

New York Times. (2009) "For N.I.H. Chief, Issues of Identity and Culture", Harris Gardner, October 5, 2009, available at: http://www.nytimes.com/2009/10/06/health/06nih.html?pagewanted=all&_r=0 (accessed July 2015)

Niederhuber, John E. (2010) "Translating Discovery to Patient Care." *JAMA* 303: 1088–1089.

Nightingale, Paul. (2000) "Economies of Scale in Experimentation: Knowledge and Technology in Pharmaceutical R&D." *Industrial and Corporate Change* 9: 315–359.

Nightingale, Paul and Martin, Paul. (2004) "The Myth of the Biotech Revolution." *Trends in Biotechnology* 22: 564–569.

Nightingale, Pail and Mahdi, Surya. (2006) "The Evolution of Pharmaceutical Innovation." In Mazzucato, Mariana and Dosi, Giovanni. (eds.) *Knowledge Accumulation and Industry Evolution: The Case of Pharma-Biotech* (Cambridge: Cambridge University Press), pp. 73–111.

NIH. (2010) *Scientific Management Review Board Report on Translational Medicine and Therapeutics,* December 2010, National Institutes of Health, available at: http://smrb.od.nih.gov/documents/reports/TMAT_122010.pdf (accessed March 2015).

NIH. (2011) "NIH Establishes National Center for Advancing Translational Sciences." National Institutes of Health News, Friday, December 23,

2011, available at: http://www.nih.gov/news/health/dec2011/od-23. htm (accessed March 2015).

NIH Office of Budget. (2011) *Actual Obligations by Institute and Center FY 1997–2010* [online], available at: http://officeofbudget.od.nih.gov/pdfs/ FY12/Actual%20Obligations%20by%20IC%201998-2010.pdf (accessed February 2012).

Novas, Carlos. (2006) "The Political Economy of Hope: Patients" Organizations, Science and Biovalue." *Biosocieties* 1: 289–305.

Novas, Carlos and Rose, Nikolas. (2000) "Genetic Risk and the Birth of the Somatic Individual." *Economy and Society* 29: 485–513.

Nussenblatt, Robert B and Marincola, Francesco M. (2013) "Emerging Concepts in Biomarker Discovery: Cancer Immunotherapy and Degenerative Diseases of the Eye as Model Systems." In Mittra, James and Milne, Christopher-Paul. (eds.) *Translational Medicine: The Future of Therapy?* (Singapore: Pan-Stanford), pp. 129–161.

OECD. (2009) "The Bioeconomy to 2030: Designing a Policy Agenda." OECD Publishing, International Futures Programme.

Ogilvie, David, Craig, Peter, Griffin, Simon, Macintyre, Sally, and Wareham, Nicholas, J. (2009) "A Translational Framework for Public Health Research." *BMC Public Health* 9: 116.

Omidvar, Omid, de Grijs, Myriam, Castle, David, Mittra, James, Rosiello, Alessandro, and Tait, Joyce. (2014) *Regenerative Medicine: Business Models, Venture Capital and the Funding Gap*, Final Project Report to ESRC, Innogen Institute, University of Edinburgh, available at: http://www.innogen.ac.uk/downloads/RegenerativeMedicine-BusinessModels-VentureCapital-and-theFundingGap-Oct14.pdf (accessed January 2015).

Orloff, John, Douglas, Frank, Pinheiro, Jose, Levinson, Susan, Branson, Michael, et al. (2009) "The Future of Drug Development: Advancing Clinical Trial Design." *Nature Reviews Drug Discovery* 8: 949–957.

Oudshoorn, Nelly. (2011) *Telecare Technologies and the Transformation of Healthcare* (Basingstoke: Palgrave Macmillan).

Pammolli, Fabio and Riccaboni, Massimo. (2000) *Consolidation and Competition in the Pharmaceutical Industry*, Paper delivered at the OHE Conference, London, October 16, 2000, Kettler, Hannah E. (ed.) (London: Office of Health Economics).

Pammolli, Fabio L, Magazzini, Laura, and Riccaboni, Massimo. (2011) "The Productivity Crisis in Pharmaceutical R&D." *Nature Reviews Drug Discovery* 19: 428–438.

Papaioannou, Theo, Wield, David, and Chataway, Joanna. (2009) "Knowledge Ecologies and Ecosystems: An Empirically Grounded Reflection on Recent Developments in Innovation Systems Theory." *Environment and Planning C: Government and Policy* 27: 319–339.

Pardridge, William M. (2003) "Translational Science: What Is It and Why Is It So Important?" *Drug Discovery Today* 18(8): 813–815.

Parry, Bronwyn. (2007) "Cornering the Futures Market in Bio-Epistemology." *Biosocieties* 2(3): 386–389.

Paul, J. E and Trueman, P. (2001) "Fourth Hurdle Reviews, NICE and Database Applications." *Pharmacoepidemiology and Drug Safety Journal* 10(5): 429–438.

Perkmann, Markus and Spicer, Andre. (2010) "What Are Business Models? Developing a Theory of Performative Representations." In Phillips, Nelson. Sewell, Graham, and Griffiths, Dorothy. (eds.) *Technology and Organization: Essays in Honour of Joan Woodward (Research in the Sociology of Organizations, Volume 29)* (Bingley: Emerald Group Publishing), pp. 265–275.

PharmaForum. (2012) "Redefining the Blockbuster Model: Why the $1 Billion Entry Point Is No Longer Sufficient—Part 1." PharmaForum Article, September 11, 2012, available at: http://www.pharmaphorum. com/articles/redefining-the-blockbuster-model-why-the-1-billion-entry-point-is-no-longer-sufficient-part-1 (accessed January 2015).

Pickersgill, Martyn. (2012) "What Is Psychiatry? Co-producing Complexity in Mental Health." *Social Theory & Health* 10(4): 328–347.

Pickersgill, Martyn. (2013) "The Social Life of the Brain: Neuroscience in Society." *Current Sociology* 61(3): 322–340.

Pickersgill, Martyn. (2014) "Debating DSM-5: Diagnosis and the Sociology of Critique." *Journal of Medical Ethics* 40(8): 521–525.

Pisano, Gary. (2006) *Science Business; The Promise, the Reality, and the Future of Biotech* (Massachusetts: Harvard University Press).

Pollock, Neil and Williams, Robin. (2010) "The Business of Expectations: How Promissory Organizations Shape Technology and Innovation." *Social Studies of Science* 40(4): 525–548.

Pols, Jeannette. (2014) "Knowing Patients: Turning Patient Knowledge into Science." *Science, Technology and Human Values* 39(1): 73–97.

Porter, Michael E. (2010) "What Is Value in Healthcare?" *The New England Journal of Medicine* 363(26): 2477–2481.

Rabeharisoa, Vololona, Moreira, Tiago, and Akrich, Madeleine. (2014) "Evidence-Based Activism: Patients', Users' and Activists' Groups in Knowledge Society." *Biosocieties* 9(2): 111–128.

Rafols, Ismael, Hopkins, Michael M, Hoekman, Jarno, Siepel, Josh, O"Hare, Alice, et al. (2014) "Big Pharma, Little Science? A Bibliometric Perspective on Big Pharma's R&D Decline." *Technological Forecasting and Social Change* 81: 22–38.

Rägo, Lembit, and Santoso, Budiono. (2008) "Drug Regulation: History, Present and Future." In van Boxtel, Chris J, Santoso, Budiono, and Edwards, Ralph I. (eds.) *International Textbook of Clinical Pharmacology*, 2nd Edition (New Jersey: Wiley), pp. 65–77.

Rai, Arti K. (2005) "Open and Collaborative Research: A New Model for Biomedicine." In Hahn, Robert. (ed.) *Intellectual Property Rights In Frontier Industries: Software and Biotech*, AEI-Brookings Press; Duke

Law School Legal Studies Research Paper Series Research Paper No. 61, pp. 31–158.

Rajan, Kaushik Sunder. (2006) *Biocapital* (Durham: Duke University Press).

Rang, H. P. (2005) "The Development of the Pharmaceutical Industry." In Rang, H. P. (ed.) *Drug Discovery and Development: Technology in Transition* (London: Churchill Livingstone), pp. 3–40.

Ratcliffe, Elizabeth, Thomas, Robert J, and Williams, David J. (2011) "Current Understanding and Challenges in Bioprocessing of Stem Cell-Based Therapies for Regenerative Medicine." *British Medical Bulletin* 100: 137–155.

Ratti, Emiliangelo and Trist, David. (2001) "Continuing Evolution of the Drug Discovery Process in the Pharmaceutical Industry." *Pure and Applied Chemistry* 73(1): 67–75.

Raunio, Hannu. (2011) "In Silico Toxicology—Non-Testing Methods." *Frontiers in Pharmacology* 2(33): 1–8. DOI: 10.3389/fphar.2011.00033.

Ray, Amrit. (2009) "Beyond Debacle and Debate: Developing Solutions in Drug Safety." *Nature Reviews Drug Discovery* 8: 775–779.

Reiss, Thomas and Hinze, Sybille. (2000) "Innovation Process and Techno-Scientific Dynamics." In Jungmittag, Andre, Reger, Guido, and Reiss, Thomas. (eds.) *Changing Innovation in the Pharmaceutical Industry: Globalization and New Ways of Drug Development* (Berlin: Springer), pp. 53–70.

Rekers, Josephine V and Hansen, Teis. (2015) "Interdisciplinary Research and Geography: Overcoming Barriers through Proximity." *Science and Public Policy* 42(2): 242–254.

Robertson, David W. Martin, Douglas K, and Singer, Peter A. (2003) "Interdisciplinary Research: Putting the Methods Under the Microscope." *BMV Medical Research Methodology* 3: 1–5.

Rose, Nikolas. (2001) "The Politics of Life Itself." *Theory, Culture and Society* 18(6): 1–30.

Rothstein, Mark A. (ed.) (2003) *Pharmacogenomics: Social, Ethical, and Clinical Dimensions* (New Jersey: John Wiley).

Rubin, Mark, A. (2015) "Make Precision Medicine Work for Cancer Care." *Nature* 520: 290–291.

Rubio, Doris McGartland, Schoenbaum, Ellie E, Lee, Linda, S, Schteingart, David E, Marantz, Paul R, et al. (2010) "Defining Translational Research: Implications for Training." *Academic Medicine* 85(3): 470–475.

Russell, William M. S and Burch, R. L. (1959). *The Principles of Humane Experimental Technique* (London: Methuen).

Saijo, Nagahiro. (2002) "Translational Study in Cancer Research." *Internal Medicine* 41: 770–773.

Salman, Rustam Al-Shahi, Beller, Elaine, Kagan, Jonathan, Hemminki, Elina, Phillips, Robert S, et al. (2014) "Increasing Value and Reducing Waste in Biomedical Research Regulation and Management." *The Lancet* 383: 176–184.

Sarkar, Indra Neil. (2013) "Translational Bioinformatics: Bridging the Biological and Clinical Divide." In Mittra, James and Milne, Christopher-Paul. (eds.) *Translational Medicine: The Future of Therapy?* (Singapore: Pan-Stanford), pp. 189–210.

Scannell, Jack W, Blanckley, Alex, Boldon, Helen, and Warrington, Brian. (2012) "Diagnosing the Decline in Pharmaceutical R&D Efficiency." *Nature Reviews Drug Discovery* 11: 191–200.

Schmid, Esther F and Smith, Dennis A. (2005) "Is Declining Innovation in the Pharmaceutical Industry a Myth?" *Drug Discovery Today* 10: 1031–1038.

Schmid, Otto, Padel, Susanne, and Levidow, Les. (2012) "The Bio-Economy Concept and Knowledge Base in a Public Goods and Farmer Perspective." *Bio-Based and Applied Economics* 1(1): 47–63.

Rial-Sebbag, Emmanuelle and Cambon-Thomsen, Anne. (2012) "The Emergence of Biobanks in the Legal Landscape: Towards a New Model of Governance." *Journal of Law and Society* 39(1): 113–129.

Sense about Science. (2013) *Evidence Based Medicine Matters,* Academy of Medical Royal Colleges, London, available at: http://www.senseabout-science.org/data/files/resources/124/Evidence-Based-Medicine-Matters.pdf (accessed March 2014).

Shickle, Darren. (2006) "The Consent Problem within DNA Biobanks." *Studies in History and Philosophy of Biological and Biomedical Sciences* 37: 503–519.

Smith, Blair H, Campbell, Harry, Blackwood, Douglas, Connell, John, Connor, Mike, et al. (2006) "Generation Scotland: the Scottish Family Health Study: A New Resource for Researching Genes and Heritability." *BMC Medical Genetics* 7(74):1–9. DOI:10.1186/1471-2350-7-74.

Snape, K, Trembath, R. C, and Lord, G. M. (2008) "Translational Medicine and the NIHR Biomedical Research Centre Concept." *Quarterly Journal of Medicine* 101: 901–906.

Soderquest, Katrina and Lord, Graham M. (2010) "Strategies for Translational Research in the United Kingdom." *Science Translational Medicine* 2(53): 1–5.

Speight, Jane. (2010) "FDA Guidance on Patient Reported Outcomes." *The British Medical Journal* 340: c2921.

Stark, David. (2009) *The Sense of Dissonance: Accounts of Worth in Economic Life* (New Jersey: Princeton University Press).

Stemerding, Dirk and Nahuis, Roel. (2014) "Implicit and Explicit Notions of Valorization in Genomics Research." *New Genetics and Society* 33(1): 79–95.

Stokes, Donald E. (1997) *Pasteur's Quadrant: Basic Science and Technological Innovation* (Washington, DC: Brookings).

Sturdy, Steve. (2012) "Looking For Trouble: Medical Science and Clinical Practice in the Historiography of Modern Medicine." *Social History of Medicine* 24(3): 739–757.

Styhre, Alexander and Sundgren, Mats. (2011) *Venturing into the Bioeconomy: Professions, Innovation, Identity* (Basingstoke: Palgrave Macmillan).

Tait, Joyce. (1990) *Biotechnology: Interactions between Technology, Environment and Society*, FAST Programme Synthesis Report No 1 Project, EU Fast.

Tait, Joyce. (2007) "Systemic Interactions in Life Science Innovation." *Technology Analysis and Strategic Management* 19(3): 257–277.

Tait, Joyce. (2009) "Upstream Engagement and the Governance of Science: The Shadow of the Genetically Modified Crops Experience in Europe." *EMBO Reports* 10: 18–22.

Tait, Joyce and Lyall, Catherine. (2007) *Short Guide to Developing Interdisciplinary Research Proposals*, ISSTI Briefing Note Number 1, available at: http://www.citsee.ed.ac.uk/__data/assets/file/0005/77603/ISSTI_Briefing_Note_1.pdf (accessed July 2014).

Tait, Joyce and Williams, Robin. (1999) "Linear Plus Model: Policy Approaches to Research and Development." *Science and Public Policy* 26(2): 101–112.

Tait, Joyce, Wield, David, Bruce, Ann, and Chataway, Joanna. (2008) *Health Biotechnology to 2030: Report to OECD International Futures Project*; "The Bioeconomy to 2030: Designing a Policy Agenda." OECD Paris, available at: http://www.oecd.org/dataoecd/12/10/40922867.pdf (accessed November 2014).

Tait, Joyce, Bruce, Ann, Mittra, James, Purves, John, and Scannell, Jack. (2014) *Independent Review on Anti-Microbial Resistance: Regulation-Innovation Interactions and the Development of Antimicrobial Drugs and Diagnostics for Human and Animal Diseases*, Report to Jim O'Neill Review on Antimicrobial Resistance, ESRC, Innogen Institute.

Terwilliger, Joseph D and Goring, Harold H. H. (2009) "Update to Terwilliger and Goring's 'Gene Mapping in the 20th and 21st Centuries': Gene Mapping When Rare Variants are Common and Common Variants are Rare." *Human Biology* 81(5–6): 729–733.

The Pharmaceutical Journal (2012) "ABPI Raises Concerns over Value-Based Pricing." *The Pharmaceutical Journal* Aug 6, 2012, available at: http://www.pharmaceutical-journal.com/news-and-analysis/news/abpi-raises-concerns-over-value-based-pricing/11104899.article (accessed July 2015).

Trusheim, Mark R, Berndt, Ernst R, and Douglas, Frank, L. (2007) "Stratified Medicine: Strategic and Economic Implications of Combining Drugs and Clinical Biomarkers." *Nature Reviews Drug Discovery* 6: 287–293.

Trusheim, Mark R, Burgess, Breon, Xinghua, Hu Sean, Lonf, Theresa, Averbuch, Steven D, Flynn, Aiden A, et al. (2011) "Quantifying Factors for the Success of Stratified Medicine: Strategic and Economic Implications of Combining Drugs and Clinical Biomarkers." *Nature Reviews Drug Discovery* 10: 817–833.

TSB. (2011) *Stratified Medicine in the UK: Vision and Road Map*, Technology Strategy Board, London. available at: https://connect.innovateuk.org/

documents/2843120/3724280/Stratified+Medicines+Roadmap.pdf/
fbb39848-282e-4619-a960-51e3a16ab893 (accessed May 2014).

Tupasela, Aaro and Stephens, Neil. (2013) "The Boom and Bust Cycle of
Biobanking—Thinking through the Life Cycle of Biobanks." *Croatian
Medical Journal* 54: 501–503.

Tutton, Richard, Kaye, Jane, and Hoeyer, Klaus. (2004) "Governing
UK Biobank: The Importance of Ensuring Public Trust." *Trends in
Biotechnology* 22(6): 284–285.

Tutton, Richard, Smart, Andrew, Martin, Paul. A, Ashcroft, Richard,
and Ellison, George T. H. (2008) "Genotyping the Future: Scientists"
Expectations about Race and Ethnicity after BiDil." *Journal of Law,
Medicine and Ethics* 36(3): 464–470.

Venter, Craig and D. Cohen, Daniel. (2004) "The Century of Biology." *New
Perspectives Quarterly* 21(4): 73–77.

Vermeulen, Niki, Parker, John N, and Penders, Bart. (2013) "Understanding
Life Together: A Brief History of Collaboration in Biology." *Endeavour*
37(3): 162–171.

Wagner, J. A, Prince, M, Wright, E. C, Ennis, M. M, Kochan, J, et al. (2010)
"The Biomarkers Consortium: Practice and Pitfalls of Open-Source
Precompetitive Collaboration." *Clinical Pharmacology and Therapeutics*
87(5): 539–542.

Waldby, Catherine. (2000) "Stem Cells, Tissue Cultures and the Production
of Biovalue." *Health: An Interdisciplinary Journal* 6(3): 305–323.

Wang, Jingsong. (2013) "Imaging Biomarkers for Innovative Drug
Development." In Mittra, James and Milne, Christopher-Paul. (eds.)
Translational Medicine: The Future of Therapy? (Singapore: Pan-Stanford),
pp. 163–187.

Weber, W. A. (2006) "Positron Emission Tomography as an Imaging
Biomarker." *Journal of Clinical Oncology* 24: 3282–3292.

Webster, Andrew. (2007) *Health, Technology and Society: A Sociological
Critique* (Basingstoke: Palgrave Macmillan).

Webster, Andrew, Haddad, Christian, and Waldby, Catherine. (2011)
"Experimental Heterogeneity and Standardisation: Stem Cell Products
and the Clinical Trial Process." *Biosocieties* 6(4): 401–419.

West, Will and Nightingale, Paul. (2009) "Organizing for Innovation:
Towards Successful Translational Research." *Trends in Biotechnology*
27(10): 558–561.

White House. (2012) *National Bioeconomy Blueprint,* available at: http://
www.whitehouse.gov/sites/default/files/microsites/ostp/national_bio-
economy_blueprint_april_2012.pdf (accessed January 2015).

Wield, David. (2013) "Bioeconomy and the Global Economy: Industrial
Policies and Bioinnovation." *Technology Analysis and Strategic
Management* 25(10): 1209–1221.

Wield, David, Hanlin, Rebecca, Mittra, James, and Smith, James. (2013)
"Twenty-First Century Bioeconomy: Global Challenges of Biological

Knowledge for Health and Agriculture." *Science and Public Policy* 40: 17–24.

Will, Catherine M. (2011) "Mutual Benefit, Added Value? Doing Research in the National Health Service." *Journal of Cultural Economy* 4(1): 11–26.

Williams, David J. (2011) "Overcoming Manufacturing and Scale-Up Challenges." *Regenerative Medicine* 6: 67–69.

Williams, Robin. (2006) "Compressed Foresight and Narrative Bias: Pitfalls in Assessing High Technology Futures." *Science as Culture* 15(4): 327–348.

Williams, Simon, Gabe, Jonathan, and Martin, Paul. (2012) "Medicalization and Pharmaceuticalisation at the Intersections: A Commentary on Bell and Figert." *Social Science and Medicine* 75(12): 2129–2130.

Wilson-Kovacs, Dana M and Hauskeller, Christine. (2012) "The Clinician-Scientist: Professional Dynamics in Clinical Stem Cell Research." *Sociology of Health and Illness* 34(4): 497–512.

Wood Mackenzie. (2004) *Executive's Guide 2004*, Biotechnology Overview, Wood Mackenzie, 2004.

Woodcock, Janet and Woosley, Raymond. (2008) "The FDA Critical Path Initiative and Its Influence on New Drug Development." *Annual Review of Medicine* 59: 1–12.

Woolf, Steven H. (2008) "The Meaning of Translational Research and Why It Matters." *JAMA* 299(2): 211–213.

Woolf, Stewart. (1974) "Editorial: The Real Gap Between Bench and Bedside." *The New England Journal of Medicine* 290: 802–803.

Woosley, R. L. Myers, R. T, and Goodsaid, F. (2010) "The Critical Path Institute's Approach to Precompetitive Sharing and Advancing Regulatory Science." *Clinical Pharmacology and Therapeutics* 87(5): 530–533.

van der Worp, Bart H, Howells, David W, Sena, Emily S, Porritt, Michelle J, and Rewell, Sarah. (2010) "Can Animal Models of Disease Reliably Inform Human Studies?" *PLOS Medicine* 7: 1–8, available at: http://www.plos-medicine.org/article/fetchObject.action?uri=info:doi/10.1371/journal.pmed.1000245&representation=PDF (accessed July 2015)

Zerhouni, Elias. (2003). "The NIH Roadmap." *Science* 302: 63–72.

Zerhouni, Elias A. (2005) "Translational and Clinical Science—Time for a New Vision." *The New England Journal of Medicine* 353: 1621–1623.

Zerhouni, Elias A, Sanders, Charles A, and von Eschenbach, Andrew C. (2007) "The Biomarkers Consortium: Public and Private Sectors Working in Partnership to Improve the Public Health." *The Oncologist* 12: 250–252.

Index